Warren Zeiller is managing curator and director of planning and development at the Miami Seaquarium. After receiving his M.A. in Business and Public Service at Michigan State and B.S. in Animal Husbandry at Colorado A. & M., he took his first post at the Seaquarium as diver and dolphin trainer. Subsequently he has served there as aquarist and student curator and curator of fishes. He is a member of the American Fisheries Society, the American Society of Ichthyologists & Herpetologists, and Coral Gables Rotary Club. He is the author of *Tropical Marine Fishes of South Florida and the Bahamas* and numerous articles in the field.

TROPICAL MARINE INVERTEBRATES OF SOUTHERN FLORIDA AND THE BAHAMA ISLANDS

GREEK GODDESS

TROPICAL MARINE INVERTEBRATES OF SOUTHERN FLORIDA AND THE BAHAMA ISLANDS

WARREN ZEILLER

Managing Curator
Miami Seaquarium®

A Wiley-Interscience Publication

JOHN WILEY & SONS

NEW YORK LONDON SYDNEY TORONTO

Phylum Mollusca
Class Gastropoda
Order Doridoidea
Family Dorididae

GREEK GODDESS*
Hypselodoris edenticulata (White)
by J. W. LaTourrette

This book was set in Electra by Baltimore Type & Composition Corporation. It was printed by Princeton Polychrome, Inc. and bound by Quinn & Boden. The designer was Jerome B. Wilke. Robert J. Fletcher supervised production.

Hypselodoris—from the Greek letter upsilon, Y, to which the shape of the brachial tufts may bear some resemblance, and Doris, a Greek sea goddess; *edenticulata*—Latin, without teeth. The specific name refers to the specimen originally described (holotype), which was found to lack denticulations on the radular teeth. Other specimens of the species since described have possessed this characteristic. Few tropical Atlantic nudibranchs discovered so far are equal in beauty to this species. The specimen depicted here is unusually large. Being heterosexual, it twice laid spirals of bright-orange fertile eggs on the aquarium glass. Eggs removed to a closed-system aquarium (recirculating water) were promptly destroyed by protozoans and other microscopic organisms. Those left in the open system (continuously running filtered seawater) developed normally; on about the tenth day, microscopic examination revealed the larvae within a frail shell within the egg case. The shell is lost prior to hatching, but verifies their molluscan heritage.
(see jacket and frontispiece)

* Common name suggested by the author.

Library of Congress Cataloging in Publication Data:

Zeiller, Warren, 1929-
    Tropical marine invertebrates of southern Florida and the Bahama Islands.

    "A Wiley-Interscience publication."
    1. Marine invertebrates—Florida—Identification.
    2. Marine invertebrates—Bahamas—Identification.
    I. Title.
QL169.Z44   592'.09'2463   74-2467
ISBN 0-471-98153-2

Printed in the United States of America

10 9 8 7 6 5 4 3 2 1

To Dianne and Todd, my daughter and son, whose innocent fascinations in all life forms are catalytic to my efforts.

ACKNOWLEDGMENTS

I wish to express my appreciation to Burton Clark, Vice President and General Manager of the Wometco Miami Seaquarium,® whose cooperation has made this book possible.

Many hours of discussion on specialized areas of research were willingly surrendered from busy schedules by graduate students of the Rosenstiel School of Marine and Atmospheric Science, University of Miami, and must be acknowledged.

For their continuing interest, encouragement, and invaluable advice in the field of marine invertebrata, I am most grateful to the following faculty members of the Rosenstiel School of Marine and Atmospheric Science: Robert Work, Research Associate; Dr. Anthony Provanzano, Adjunct Professor; Dr. Lowell Thomas, Associate Professor; Dr. Harding Owre, Professor; and Dr. Frederick M. Bayer, Professor.

CONTENTS

TROPICAL MARINE INVERTEBRATES OF SOUTHERN FLORIDA AND THE BAHAMA ISLANDS

INTRODUCTION

My first book, *Tropical Marine Fishes of Southern Florida and the Bahama Islands*, was designed as a series of color photographs representative of fishes to be found from estuary to coral reef. Such a format should serve a broad audience as a visual identification guide to healthy living specimens. It also reduces the need to include the usual systematic data that must accompany line drawings or artists' renditions of specimens. In the present book the same basic aim is directed to the lower life forms, the abundance of which is phenomenal. In studying vertebrates, a small reef the size of one's living room would be found to be the ecological habitat of hundreds of fishes representing several handfuls of genera and species. A random square surface meter of the same reef would contain invertebrates in such abundance that their number would undoubtedly surpass that of the fish population for the entire reef. Keep in mind that plant life is not included in our hypothetical reef study. Furthermore, we overlook the untold numbers of microfauna not discernible to the naked eye, such as protozoans. We can even dissuade ourselves from becoming engrossed in the nearly 5000 sponges, Porifera.

With a few exceptions, the tropical marine invertebrates depicted here are those commonly found throughout the shallow waters of southern Florida and the Bahama Islands. The known geographic range of many of these species might extend as far as from Cape Cod southward through the West Indies and Gulf of Mexico to Brazil. A few are confined to a fairly specific locale. The title of this volume delineates the range within which the specimens were collected. Broader geographic parameters have been avoided, as have weights or measures of species. Such figures generally are presented as maximums or averages and are subject to constant revision as new data are compiled. This type of information is important in marine biological studies, but is not essential to this work.

In short, our observations are confined to representative samples of half a dozen of the 27 invertebrate phyla, the creatures most often encountered by collectors (whether professional or lay), students, scientists, scuba or free divers, or just sometime weekend waders. More than likely, the creatures encountered will be endemic to beach or shallow-water environments. One might be a ghost crab seen browsing for an early morning snack among debris washed ashore during the last high tide. Chances are the view will be brief; approached to within a yard or two, it will skitter away and disappear into its sand burrow. A wealth of shells are left stranded by that same receding tide, but the malacologist must be an early-bird. Other beachcombers, animals, surf, and the returning tide all vie for the collector's prize. Early evening can be rewarding as well. Gastropods emerge from their hiding places to search for food near the shore. One need not fear stingrays or other potentially hazardous creatures that frequent the shallows at this time for the same reason. Comfortable footwear will afford some protection, and a shuffling gait through the water will stir up the undesirables and put them to flight ahead of the collector. Hermit crabs are often tucked away in stranded shells, patiently awaiting rescue at the return of the tide. Other crabs are in the rocks or wherever; they are so abundant and diverse in habitat that they are encountered just about anywhere. During daylight dozens of species of noctural invertebrates are secreted under every stone or piece of wood. However, one should never fail to return overturned objects to their original positions after collections and observations are completed. It seems a small thing to ask when one considers that exposed creatures will rapidly succumb to predation and entire colonies will be lost.

Normally beyond the beachcomber's reach, sodden sponges washed ashore during last night's storm will yield an abundance of life. When the tissues are gently separated or the sponge is cut into pieces, crabs, shrimps, shells, worms, and others will be revealed in every large cell and passage (condominium living is old-hat in the sea). Floats of sargassum weed in the open sea or washed ashore provide another unique habitat highly specific in its life forms. All manner of small fishes are attracted to these communities wherein they feed on the invertebrate inhabitants: sargassum crabs, sargassum shrimps, and others generally too small to "eat back." Sustained windy weather earns the disgust of boatmen and swimmers. The former forget or run out of their Dramamine and fall to the ravages of mal-de-mer; the latter dislike entering the water amid the irritant and odoriferous debris. Windrowed on the beach, the wind- and water-borne mass becomes a breeding ground for hordes of repugnant insects. Let others shy away from the shore at such times. This is when the beachcomber will reap a harvest of pelagic creatures. Stinging Portuguese men-of-war litter the sand, with gas-distended floats twisting this way and that, as if trying to catch an offshore breeze to carry them back to sea. The sun's strong rays soon quiet their struggles and bleach rich blue hues to lifeless translucent tissue. Most often they are accompanied by *Velella*, the by-the-wind sailor. Feeding on *Velella* and now carried ashore with it may be several species of lovely purple sea snails and possibly the tiny feathered nudibranch *Fiona*.

Inch-long, delicate, white, coiled shells (not depicted photographically in the text) will have survived the tortuous journey through the surf. They are the vestigial remnants of *Spirula*, a deep-water offshore squid that displays its presence after death by casting its chambered internal shell upon the shore. Driftwood, coconuts, old bottles, and other flotsam carry in numerous molluscan forms, including nudibranchs—as well as barnacles, gooseneck barnacles, bristle worms, and flatworms, to mention just a few. If small enough, a drifting colony like this makes a fascinating transplant to an aquarium. However, chances are its survival will be brief; it is difficult to maintain alone, and in a captive environment will soon be picked clean by the fishes.

Thus, although the native habitat of many life forms included in this volume normally may be far offshore or in deeper waters, each has been taken within man's easy grasp. For this reason I have not restricted them to some other book. The reader is just as apt to find them near at hand regardless of where they are meant to be (their specific environmental niche) and will want to satisfy his curiosity as I have.

All the specimens included here were photographed live in aquaria by Seaquarium staff photographers over a period of some 15 years. Three exceptions photographed in the sea are the corals *Acropora*, *Diploria*, and *Mussa*. Rather than present creatures in their ecological niche, the intent is to display as many gross features as is possible for each species. The inventiveness of the photography artists has accomplished this end through numerous means. A large reflector concentrating sunlight on a water-filled petri dish enhanced the brilliant hue of the tiny lima shrimp that pranced within it. An illusion of distant reef was effected by placing a color photograph behind an aquarium. A sheet of frosted acetate between subject and background slightly diffused the photograph and further accentuated the distance factor. An aquarium with a backplate of ground glass serves as a rearview screen on which are projected suitable color transparencies that simulate the natural environment of a particular species. There are times when no more than a simple colored board is deemed a suitable background in order to prevent distraction from the subject. Innovations continue and patience is endless; for these reasons, each photograph is credited individually. The brief captions accompanying each photograph(s) include three distinct divisions.

COMMON NAME —————————————— Uniformity and standardization in the area of common names are lacking. The vernacular names employed are those gleaned from texts or used by persons knowledgeable in the field. Often common names are nonexistent. In these cases I have taken the liberty of suggesting one, cautiously

3

marked with an asterisk and the phrase "Common name suggested by the author." An eyebrow or two will surely be raised by some workers or old salts who recognize the given specimen as thus or so. However, the names devised are offered in good faith, having been derived from generic or specific nomenclature where possible. In a few cases, where the scientific nomenclature fails to inspire the vernacular, I have tried to utilize adjectives descriptive of the gross appearance of the species. Rules established by the American Fisheries Society for determining the common names of fishes should apply to invertebrates as well. Call them what you will, the only reasonably stable system on a worldwide basis is that of scientific binomial nomenclature.

SCIENTIFIC NAME _____ Scientific names are constant and serve as a common denominator in any language. They place each specimen in a specific and generic niche within the family, order, class, and phylum. Systematic lineage often employs the additional divisions of subspecies, subgenus, subfamily, suborder, and subclass. Except for the subspecies *Cassis madagascariensis spinella*, *Lima scabra tenera*, and *Bursatella leachi plei*, all subdivisions are avoided for the sake of brevity.

Generic and specific nomenclature is given on each page. The photographs, however, are preceded by the specimen's phyletic lineage from phylum through family. Throughout my research for this book I have found that only generic and specific names are used in describing a particular creature. In order to determine the family to which it belongs, one must thumb through the text in reverse until the family name is encountered. The same holds true for the order, class, and so on. Heaven help the serious student who in the course of reverse thumbing misses the sequential family name and takes the next one in error (the same applies to order and class). Thus I have listed the phyletic lineages in order to eliminate error and render this information available at a glance.

ETYMOLOGY _____ The terms used in zoological nomenclature are rooted in many languages, principally Latin and Greek. Analysis of these terms promotes understanding and should eliminate the need to memorize them. Some names are so cleverly conceived as descriptions of genera and species that they are unforgettable. However, several common practices serve to cloud the function of descriptive nomenclature: the use of anagrams and patronyms. The latter is understandable in that systematists enjoy honors accorded them by their fellows. The former is less defensible: the creation of anagrams by rearranging the letters of existing terms serves only to confuse.

At this stage of aquarium technology it is easier to portray hundreds of species of marine invertebrate forms in a book than to display them alive. The word "alive" must be differentiated from holding them for a time until they die of starvation or other causes. Rather, they must be able to continue life as in the sea itself by being provided with a reasonable simulation of their natural environment. Achievement of this in closed-system aquariums like those maintained by hobbyists and inland oceanariums is difficult. Even such open systems as those found in installations fringing the ocean (e.g., the Seaquarium) have their share of problems in spite of the continuous pumping of filtered seawater through their aquariums.

The seas in the great basins of the earth are stable environments for the life contained therein. Transfer any segment regardless of size or volume to an aquarium and the stability is lost. From the very first minute that our hypothetical oceanic segment is contained many things begin to happen. Each change renders life more difficult and in time, intolerable for marine organisms. A number of basic aquarium practices aid in stabilizing the variables: filtration, aeration, a limited number of chemical treatments, and so on. Data on these are available in a host of "how to do it" publications. There are, perhaps, fewer variables in open systems wherein seawater is continuously filtered, passed through the artificial environment, and returned to the sea ostensibly never to be used again. Even here changes do occur, and the seawater cannot be construed as being "natural."

No less important than variables in water parameters are considerations of foodstuffs for the contained marine organisms. This always seems to be the next order of business once a creature has become captive. Our moral and aesthetic judgments seldom permit predation or similar natural forms of food gathering. We persist in attempting to develop other than natural foods to answer this need. This is a perfectly acceptable course of action that has proved to be valid in other forms of animal husbandry. Most fishes, in fact, can be made to thrive on artificial diets. However, the great multitude of lower life forms are more difficult to satisfy. Some may simply browse on the waste of other creatures. Others are highly specific in dietary requirements: one species must subsist at the expense of another, and no other will serve as an acceptable substitute. Thus until a highly specific artificial diet is devised the first species cannot survive without the culture of its food organism as well. To compound the issue, the food organism might be a plant form, and except for a limited number of algal forms, maintenance of plant life in the aquarium has yet to be mastered.

Let us not dwell further on that which has not been accomplished to date. Herein lies the fascination of the marine aquarium.

Every day involves some new challenge, and the introduction of each new species into the aquarium involves its own specific set of "unknowns." They will be conquered one by one. More than likely, as has been so often the case in the past, the hobbyist with his infinite patience and continuing loving care will add as much to our knowledge as anyone. At least the continuously increasing demands of the home aquarist on design engineers and manufacturers will stimulate research and development to the same end. Scientists, technicians, and students account for tremendous advances in the field of marine aquariology. Field studies are difficult at best and are often impossible in the sea. In a broad variety of situations the object of scrutiny must be brought to the laboratory for observation and testing. The order of first magnitude, then, must be to keep the specimens *alive*; thereafter studies can proceed. A third major interested group is comprised of the oceanariums. Every year representatives of these large public and private institutions meet at a symposium in order to exchange new data. Furthermore, books as well as technical and trade journals do much to augment the information available to all in the field.

Perhaps now you will realize why this is definitely not a "how to do it" book. Although a relatively small number of invertebrate species are held for varying lengths of time, no more than a handful are carried through anything resembling a life cycle; with few exceptions they are of commercial importance and so have earned financial backing for investigative scrutiny. Sooner or later the rest will become subjects of interest to scientists, students, or aquarists. Patient study, a little luck, and originality in thought and purpose will surely yield success. Certainly if the comments, information, and photographs in this book stimulate interest and action, my purpose will have been achieved.

PHYLUM COELENTERATA
THE POLYP ANIMALS

The class Hydrozoa (hydralike animals) is comprised of hydroids, millepores, and hydrocorals. Scyphozoa (cup animals) is a class containing the common jellyfishes and sea nettles. The class Anthozoa (flower animals) is represented by the sea anemones, stony corals, soft and horny corals, and sea pens. All are contained within the fascinating phylum Coelenterata and exhibit some common characteristics: each is constructed of well-defined tissues, has a digestive tract that opens to a mouth but lacks an anus, and is of distinct form and symmetry. Except for a few hydrozoans, the members of this phylum are exclusively marine. Characteristic of coelenterates, and unique to them, are specialized cells called nematocysts. There are nematocysts that adhere to effect certain modes of transportation or affix to prey; others entwine projections to hold small creatures fast, the most important being those capable of piercing prey (including humans) and injecting a virulent toxin. In most cases humans are insensitive to these, but some must be given wide berth in order to escape painful injury, examples being stinging corals and the Portuguese man-of-war. Of additional interest is the point that certain mollusks, nudibranchs in particular, feed on nematocyst-bearing coelenterates and store the undischarged cells within their own cerata for defense (see *Glaucus marinus*). With the exception of anemones, success in maintaining polyp animals in aquariums has been limited. Preliminary investigations with stony corals have achieved a reasonable level of success, enough to warrant a display of most of those depicted on the following pages. Specimens of *Montastraea* and *Mussa* have fared particularly well in a separate display illuminated by long-wave ultraviolet fluorescent light *alone*.

Phylum Coelenterata
Class Hydrozoa
Order Hydrocorallinae
Family Milleporidae

FIRE CORAL
Millepora alcicornis Linnaeus
by J. W. LaTourrette

Millepora—words in Latin for a thousand and Greek for soft stone allude to innumerable polyps within the colony; *alcicornis*—a Latin term for horned, referring to the general shape assumed by this encrusting coral, as seen here on the upper valve of *Spondylus americanus*. Skin contact will result in a painful irritation.

9

STINGING CORAL
Millepora complanata Lamarck
by J. W. LaTourrette

Millepora—Latin for a thousand and Greek for soft stone; *complanata*—flattened in Latin. Fluted mustard-colored columns warns against physical contact. When touched, will leave a slimy substance adhering to the skin. Contact with body areas less heavily calloused than the palms of the hands will produce an angry rash of lasting duration.

Phylum Coelenterata
Class Hydrozoa
Order Siphonophora
Family Chondrophoridae

PORTUGUESE MAN-OF-WAR
Physalia pelagica Linnaeus
by J. W. LaTourrette

Physalia—a combination of Greek and Latin words meaning bellows and wing, alluding to the gas-filled float that changes shape to catch the prevailing wind; *pelagica*—of the sea, Greek. Various species of fishes, *Nomeus gronowi* in particular, find refuge among the dangerous tentacles. They are safe only as long as they avoid the elastic mass. The man-of-war is responsible for injury to swimmers. To render first aid, remove tentacles adhering to the skin, wash thoroughly with soap and water (do not rub with sand), and apply mentholated calamine lotion, man-of-war salve, or a solution of meat tenderizer for relief. Severe cases may require treatment for shock. Beached men-of-war are every bit as dangerous as live ones swimming free.

Phylum Coelenterata
Class Hydrozoa
Order Siphonophora
Family Chondrophoridae

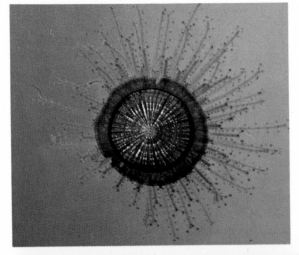

PORPITA
Porpita umbella (Muller)
by Edward T. LaRoe*

Porpita—derived from a Greek word for pin or brooch; *umbella*—Latin for an arrangement springing from a common center and forming a flat or rounded cluster. Tentacles of the dime-sized specimen are extended for feeding. Sea snails, blue glaucus, and others, in turn, find them a tasty meal.

* Rosenstiel School of Marine and Atmospheric Science, University of Miami.

BY-THE-WIND SAILOR
Velella velella (Linnaeus)
by Mike Davis

Velella—generic and specific names are derived from the Latin for under sail. One of the floating hydrozoan colonies composed of a single central large-mouthed feeding tube beneath a gas-filled float. The float is surrounded by rows of reproductive bodies; its circumference is armed with stinging tentacles, which are innocuous to man. *Velella* are fed upon by purple sea snails and nudibranchs found on them, and, when in great shoals, by the huge ocean sunfish.

Phylum Coelenterata
Class Scyphozoa
Order Rhizostomae
Family Cassiopeidae

12

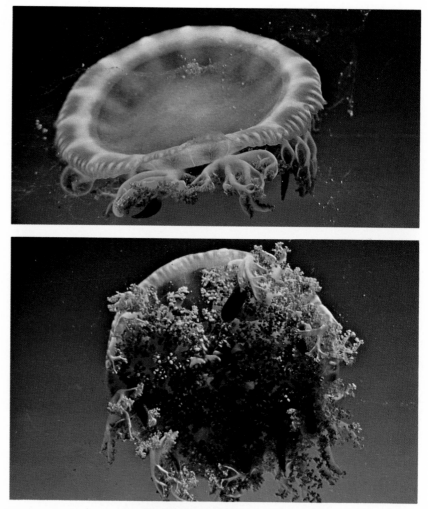

JAMAICAN CASSIOPEA*
Cassiopea xamachana Bigelow
by Don Renn

Cassiopea—Latin for helmet-shaped; *xama-chana*—from an Indian word for the island of Jamaica. Commonly found along the shore and among tangles of mangrove roots. When not swimming with a pulsing movement of the bell they assume an upside-down feeding position. Brownish-green colors are reflected from algae living within the tissues. In the aquarium without proper nourishment and/or bright sunlight they diminish slowly in size and die. Sea turtles were reared in our laboratory on a diet of *Cassiopea* and proved very vigorous and healthier than others on carnivorous or omnivorous diets!

* Common name suggested by the author.

Phylum Coelenterata
Class Scyphozoa
Order Semaeostomeae
Family Pelagidae

COMMON SEA NETTLE
Chrysaora quinquecirrha Agassiz
by J. W. LaTourrette

Chrysaora—from Greek words for golden and ugly (perhaps in these creatures Agassiz saw something beautiful and at the same time ugly); *quinquecirrha*—from the Latin word for five, referring to the number of tentacles between each successive pair of sense organs, and another Latin word for curl or tendril. Younger specimens are more colorful, but all are armed with warty clusters of nematocysts scattered over the exumbrella. Sea nettles like this range from New England southward to Bermuda, Florida, and Brazil. In some areas (Chesapeake Bay) their great numbers are a hazard to bathers and a hindrance to tourism.

13

Phylum Coelenterata
Class Anthozoa
Order Zoanthidea
Family Zoanthidae

14

GREEN SEA MAT
Zoanthus sociatus (Ellis & Solander)
by J. W. LaTourrette

Zoanthus—from Greek words for life and flower, probably alluding to a plantlike animal form; *sociatus*—Latin for social, growing in a community. Small anemone forms that colonize dense areas. Those shown here are in a closed position.

Phylum Coelenterata
Class Anthozoa
Order Actinaria
Family Aliciidae

STINGING ANEMONE
Lebrunia danae (Duchassaing & Michelotti)
by Don Renn

Lebrunia—Latin, meaning very brown; *danae*—a patronymic name. As implied by the common name, this species is particularly virulent. The appearance of *Lebrunia* is distinctive due to the variety of shapes of the tentacles. Often they are so well camouflaged among algae and other growths that they elude the most experienced collector.

Phylum Coelenterata
Class Anthozoa
Order Actinaria
Family Actiniidae

MAROON ANEMONE
Actinia bermudensis Verrill
by J. W. LaTourrette

Actinia—from the Greek word for ray; *bermudensis*—of Bermuda. The specific name refers to the locale from which the species was first described and must not be construed as the range of geographic distribution. (This will hold true for all such nomenclature.) Most anemones will survive for a time in aquariums. They require little care and need be fed once or twice a week. Waste is disgorged from the single opening in a slime-coated bolus and should be removed from the system at once to prevent fouling.

Phylum Coelenterata
Class Anthozoa
Order Actinaria
Family Actiniidae

16

SEA ANEMONE
Condylactis gigantea (Weinland)
by J. W. LaTourrette

Condylactis—Greek words for knuckle and ray describe the complete mobility of the tentacles; *gigantea*—giant in Greek. Graceful tapered arms waving gently in ocean currents belie the stinging cells within. Edible creatures of any size capable of being restrained are rendered helpless and consumed. Unlike corals, these polyp animals are motile and move about or anchor with a basal disk. At least one species of Pacific clownfish (*Amphiprion sebae*) is known to accept this Atlantic actinarian as host.

Phylum Coelenterata
Class Anthozoa
Order Actinaria
Family Stoichactiidae

SUN ANEMONE
Stoichactis helianthus (Ellis)
by J. W. LaTourrette

Stoichactis—a row of rays; *helianthus*—the sun flower: four Greek words are combined in the generic and specific nomenclature. One of the more virulent anemones, it will stick to anything that touches it. *Stoichactis* grows to a large diameter and reminds one of Pacific anemones commonly marketed for the aquarium trade; both serve as hosts to clownfishes in the classical behavioral arrangement known as mutualistic symbiosis.

Phylum Coelenterata
Class Anthozoa
Order Actinaria
Family Phymanthidae

RED BEADED ANEMONE
Phymanthus crucifer (Lesueur)
by J. W. LaTourrette

Phymanthus—from the Greek word for tumor; *crucifer*—cross in Latin. This species is highly variable in form, as manifested by the two specimens shown here. The clearly lined specimen to the right is the most handsome of its kind I have found in Florida waters.

Phylum Coelenterata
Class Anthozoa
Order Actinaria
Family Hormathiidae

HERMIT ANEMONE*
Calliactus tricolor (Lesueur)
by Peter Vila

Calliactus—Greek for beautiful rays; *tricolor*—three-colored. A symbiont often found in relationship with certain species of hermit crabs, the latter providing rapid transit for the more sessile hitchhiker. When the host dies or moves to a larger shell, *Calliactus* transfers as well. What fascinating mechanisms in crab or anemone triggers the move? Color variations of the species are shown above.

* Common name suggested by the author.

18

Phylum Coelenterata
Class Anthozoa
Order Actinaria
Family Aiptasiidae

PALE SEA ANEMONE
Aiptasia pallida (Verrill)
by Don Renn

Aiptasia—from Greek for stretching; *pallida*—pale in Latin. Small, long-stalked actinarians, found in abundance in shallow waters; they pass through Seaquarium water systems and bloom throughout aquaria. These are clustered on a West Indian top shell, *Livona pica*.

19

RINGED ANEMONE*
Bartholomea annulata (Lesueur)
by J. W. LaTourrette

Bartholomea—a patronymic name; *annulata*—Latin for ringed. This delicate anemone reacts to stimuli with startling alacrity. It is known to be a symbiont with certain mycid shrimp and *Alpheus* species of pistol or snapping shrimps.

* Common name suggested by the author.

Phylum Coelenterata
Class Anthozoa
Order Scleractinia
Family Acroporidae

STAGHORN CORAL
Acropora cervicornis (Lamarck)
by J. W. LaTourrette

Acropora—a hill top or peak of soft stone in Greek;
cervicornis—from two Latin words for tawny and horned.
Frail stands develop over a period of many years, their
beauty as part of the living reef almost beyond compre-
hension. Collectors, working from boats with clam rakes,
can reduce nature's work to rubble in a matter of minutes.

20

ELKHORN CORAL
Acropora palmata (Lamarck)
by J. W. LaTourrette

Acropora—a Greek term referring to a peak of soft stone;
palmata—Latin for broad, as in the palm of the hand;
in this instance as in the broad rack of the great American
elk. A diving partner of mine once thrust his foot upon
this coral, cutting the heel to the bone. He suffered
periodic eruptions of angry boils under the arms and in
the groin for several years thereafter. Massive doses of
vitamin C effected relief—time, the only cure.

Phylum Coelenterata
Class Anthozoa
Order Scleractinia
Family Poritidae

YELLOW PORITES*
Porites astreoides Lamarck
by Peter Vila

Porites—from Greek meaning full of holes, or porous; *astreoides*—starlike in Greek. Tiny polyps reveal only a bright yellow mat to the naked eye, but display their true form under a powerful hand lens.

* Common name suggested by the author.

21

CLUBBED FINGER CORAL
Porites porites (Pallas)
by J. W. LaTourrette

Porites—both the generic and specific names have to do with pores or passages; they are rooted in Greek. Structural differences in stem diameter and branch ends (whether swollen as shown here or not) help classify numerous closely related species of the genus. *Porites* specimens collected in Biscayne Bay often survive for more than a year in our live coral display.

Phylum Coelenterata
Class Anthozoa
Order Scleractinia
Family Faviidae

BRAIN CORAL
Diploria labyrinthiformis (Linnaeus)
by J. W. LaTourrette

Diploria—Greek for twofold; *labyrinthiformis*—a Latin word, derived from Greek, meaning labyrinthine in form, in reference to a structure formed of many winding passages. This formation is not particularly large for the species, but is too massive to be moved by one man. Slabs of fossilized coral from ancient reefs are used today for facing modern building walls. The front sidewalk of my home was created from this material more than 30 years ago; labyrinthian patterns on its surface are formed from brain corals like the formation shown here.

22

COMMON BRAIN CORAL
Diplora strigosa (Dana)
by J. W. LaTourrette

Diploria—Greek for dual or twofold; *strigosa*—furrowed or channeled in Latin. These specimens are particularly colorful for this small species and have proved hardy in an open-system aquarium.

Phylum Coelenterata
Class Anthozoa
Order Scleractinia
Family Faviidae

COMMON ROSE CORAL
Manicina areolata (Linnaeus)
by Peter Vila

Manicina—hand-shaped in Latin; *areolata*—with small spaces, Latin. The swollen white tissue of the specimen to the right has lost its algae cells during confinement. The rose coral is an excellent food for butterflyfishes and other reef dwellers.

LARGE STAR CORAL
Montastraea cavernosa (Linnaeus)
by Peter Vila

Montastraea—from the Latin for mountain and Greek for star; *cavernosa*—Latin for cavernous, full of hollows. One of the larger polyped reef-building corals that reflect emerald green under ultraviolet light.

Phylum Coelenterata
Class Anthozoa
Order Scleractinia
Family Oculinidae

IVORY BUSH CORAL
Oculina diffusa Lamarck
by Peter Vila

Oculina—Latin for budded; *diffusa*—Latin for diffused, or spread out (branched). This species has proved particularly hardy in the open-system aquarium. It plays host to numerous serpulid worms and assorted mollusks. Attached bivalve mollusks did not survive in this display. Long after, when even the coral polyps had died, the serpulid worms continued to thrive. Their small, feathery, colorful crowns continued to pop in and out of the protective tube homes, busy with whatever it is that serpulid worms do.

24

Phylum Coelenterata
Class Anthozoa
Order Scleractinia
Family Trochosmiliidae

BRAIN CORAL
Meandrina meandrites (Linnaeus)
by J. W. LaTourrette

Meandrina—both scientific names refer to the surface convolutions of the colony, having been taken from the Greek-derived Latin word for winding (meandering). Many of the massive corals are formed with surface patterns so similar that it is nearly impossible to identify them without scrutinizing the skeletal structure of individual polyps. Thus most bear the simple common name "brain coral."

25

DOMED STAR CORAL*
Dichocoenia stokesi Edwards and Haime
by Peter Vila

Dichocoenia—from the Greek words for two and common; *stokesi*—a patronymic name. The small domes are composed of conspicuously oval corallites, making identification of living colonies an easy task in comparison to other coralline forms such as brain corals.

* Common name suggested by the author.

Phylum Coelenterata
Class Anthozoa
Order Scleractinia
Family Trochosmiliidae

PILLAR CORAL
Dendrogyra cylindricus Ehrenberg
by J. W. LaTourrette

Dendrogyra—Greek terms alluding to a round tree or stick; *cylindricus*—cylindrical in Latin. Erect knobbed pillars achieve great size; the 10-inch stand shown here has barely started. I have seen magnificent stands with multiple spires soaring 10 feet or more from reef to the surface of the sea. Divers often refer to these as "pipe organ corals." One would have to search far for a more breathtaking sight.

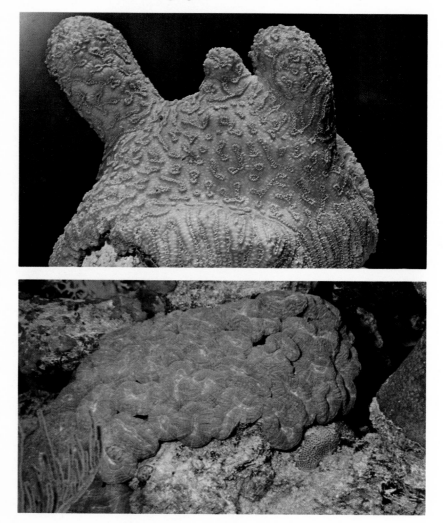

26

Phylum Coelenterata
Class Anthozoa
Order Scleractinia
Family Mussidae

LARGE FLOWER CORAL
Mussa angulosa (Pallas)
by J. W. LaTourrette

Mussa—from the Latin word for murmur or speak softly, perhaps in reference to the soft expanded tissues, as seen here; *angulosa*—angular in Latin. Note the apparent lack of tentacles, which are obvious to the naked eye in most coralline forms. This species fluoresces beautifully under ultraviolet illumination.

Phylum Coelenterata
Class Anthozoa
Order Scleractinia
Family Caryophillidae

FLOWER CORAL
Eusmilia fastigiata (Pallas)
by J. W. LaTourrette

Eusmilia—Old Greek words meaning true or nice and *smilax*, a name for plants now restricted to a genus of lily; *fastigiata*—sloping or gabled in Latin. Anemonelike polyps expand to feed on the dextrose solution introduced into the open-system aquarium. Others remain closed, displaying their stony cups.

27

Phylum Coelenterata
Class Anthozoa
Order Corallimorpharia
Family Actinodiscidae

FALSE CORAL
Ricordia florida Duchassaing & Michelotti
Cluster by Peter Vila, singles by J. W. LaTourrette

Ricordia—a patronymic name; *florida*—flowery in Latin.
The recently collected green specimens reflect algae
(zoanthelli) within their tissues; the brown specimen
has lost these during confinement in the indoor aquarium.
Although the general appearance would lead one to believe that more than one species is depicted, such is not
the case. Note that all protuberances on the disk are
rounded while those of *Rhodactis* bear several small
points each.

28

Phylum Coelenterata
Class Anthozoa
Order Corallimorpharia
Family Actinodiscidae

RED FALSE CORAL*
Rhodactis sanctithomae (Duchassaing & Michelotti)
by Peter Vila

Rhodactis—from two Greek words for red and ray; *sancti-thomae*—of St. Thomas, where the holotype (first specimen described) was probably collected. This species has survived longer than *Ricordia* in an artificial environment. When fully expanded they display pale-blue lines from center to margin; these may be the rays to which the generic name refers.

* Common name suggested by the author.

29

PHYLUM PLATYHELMINTHES
THE FLATWORMS

There are three classes of flatworms, of which two are parasitic: the flukes (Trematoda) and tapeworms (Cestoda). The third class (Turbellaria) consists of five orders of free-living flatworms. One, Polycladida, is almost exclusively marine and is here represented by two species of the family Pseudoceridae. One of these is unidentified as to species, for much remains to be done in the systematics of polyclad worms.

Incredibly, within a few brief sentences I have by-passed more than 9000 species known within this phylum.

Phylum Platyhelminthes
Class Turbellaria
Order Polycladida
Family Pseudoceridae

POLYCLAD FLATWORM
Pseudoceros sp.
by J. W. LaTourrette

Pseudoceros—Greek for false and Latin for wax, the latter alluding to its texture; unidentified as to species. This 1-inch specimen emerged from a live sponge and promptly laid rows of tiny, hard, white eggs on the aquarium glass. Despite our efforts they failed to hatch, and the flatworm did not survive long in the captive environment.

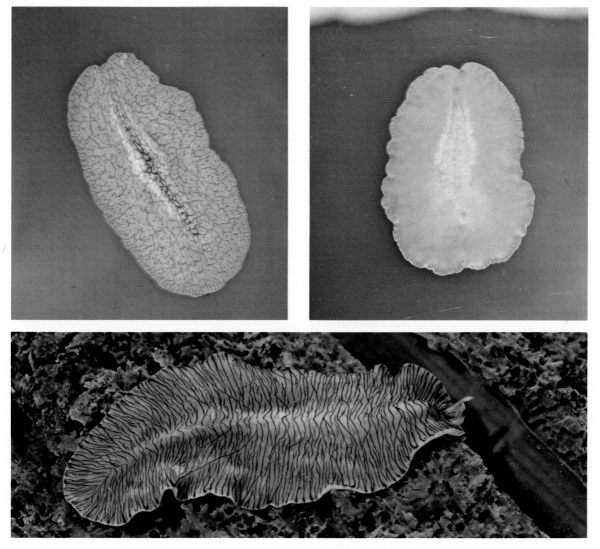

31

CROZIERS FLATWORM*
Pseudoceros crozieri Hyman
by Peter Vila

Pseudoceros—Greek and Latin words for false wax; *crozieri*—a patronym. Hornlike sensory organs are clearly displayed on the anterior dorsal surface of this handsome species.

* Common name suggested by the author.

PHYLUM MOLLUSCA
THE MOLLUSKS

The phylum Mollusca is second only to the arthropods in number and is represented by more than 40,000 species. It is particularly diverse, as displayed by the specimens in the following section that include four of the six molluscan classes. In all probability, the most primitive is Monoplacophora, known only from fossil records until the discovery of living specimens a dozen years ago. These tiny, saucer-shaped, limpetlike shells were taken from a depth of 5000 meters off the Pacific coast of Mexico. This is one of the classes not represented here. The class Amphineura encompasses the coat-of-mail shells, or chitons, together with two groups of aberrant shell-less forms. The largest and most diverse class is *Gastropoda*. Spirally coiled snails, conical or flat-shelled limpets, nudibranchs, terrestrial snails, and slugs are placed within this group. A small class typified by tusk shells is Scaphopoda; it contains only a few genera, which to date have escaped our photographic record. A much larger group and the most valuable commercially is Pelecypoda. Clams, oysters, mussels, and other two-valved shells are included therein (another class name gaining popularity is Bivalvia). The most active and specialized of mollusks are in the class Cephalopoda. These creatures may bear a chambered shell, as in *Nautilus*; an internal shell, as in squids and cuttlefishes; or no shell at all, as in octopods.

Few marine invertebrate forms have enjoyed the general popularity of mollusks. Their handsomely sculptured shells have served man as prized possessions throughout the course of history. Although other invertebrates are found in fossil records, even the watery jellyfishes, only shelled mollusks (and stony corals) create their own everlasting memorial.

Phylum Mollusca
Class Gastropoda
Order Archaeogastropoda
Family Fissurellidae

CAYENNE KEYHOLE LIMPET
Diodora cayenensis (Lamarck)
by J. W. LaTourrette

Diodora—from the Greek *Dios*, the genitive of Zeus, and *dora*, hide, or skin; *cayenensis*—a geographic reference (also an allusion to cayenne pepper, a more pointed description of the pigmentation of the dorsal surface of the soft body parts). The small opening at the point of the shell is the "keyhole" alluded to in the vernacular name.

33

CANCELLATE FLESHY LIMPET
Lucapina suffusa (Reeve)
by J. W. LaTourrette

Lucapina—etymology obscure, possibly from the Latin words *luc* for shining or conspicuous and *pinna* for winged or feathered in allusion to the frilly scarlet tissues; *suffusa*—Latin for spread out, or suffused. The muscular foot both transports and firmly anchors this univalve. When threatened, mantle and anal funnel at the apex withdraw beneath the conical shell.

Phylum Mollusca
Class Gastropoda
Order Archaeogastropoda
Family Turbinidae

STAR SHELL
Astraea caelata (Gmelin)
by Peter Vila

Astraea—derived from the Greek word for star; *caelata*—engraved in relief, Latin. The heavy, nearly hemispherical operculum affords impenetrable protection when withdrawn into the mouth of the shell, exposing only a small area of its flat surface.

Phylum Mollusca
Class Gastropoda
Order Archaeogastropoda
34 **Family** Neritidae

BLEEDING TOOTH
Nerita peloronta Linnaeus
by J. W. LaTourrette

Nerita—a Latin name for a sea snail; *peloronta*—etymology obscure; possibly from a Greek term for clay or mud, or a reference to a dusky or dark color. (The specific name may also have been derived from the Latin-derived word *peloric*, an adjective descriptive of abnormally regular or symmetrical form, as in a flower.) This species is the more colorful of half a dozen family members found along rocky shores. It is characterized by the blood-red parietal area that bears one or two white teeth.

Phylum Mollusca
Class Gastropoda
Order Archaeogastropoda
Family Janthinidae

COMMON PURPLE SEA SNAIL
Janthina janthina Linnaeus
by J. W. LaTourrette

Janthina—both the generic and specific names are derived from a Greek word for violet-colored. A bubble raft produced by the snail keeps it afloat and within reach of food among the tentacles of *Velella, Porpita,* and possibly *Physalia* with which it drifts across the oceans. The raft also supports the mass of pink eggs. Sustained winds drive all of the above ashore, to the dismay of bathers, but to the delight of the beachcomber.

Phylum Mollusca
Class Gastropoda
Order Archaeogastropoda
Family Strombidae

FLORIDA FIGHTING CONCH
Strombus alatus Gmelin
by J. W. LaTourrette

Strombus—top in Greek, in allusion to the spiral shell; *alatus*—Latin for winged, a reference to the lip of the shell, which has not achieved the massive proportions in the younger specimen on the right. This is a fairly common trophy taken along the shoreline. The adult specimen is utilizing its pointed operculum to right itself. Live hand-held specimens will flail about wildly in the same manner.

Phylum Mollusca
Class Gastropoda
Order Archaeogastropoda
Family Strombidae

QUEEN CONCH
Strombus gigas Linnaeus
by J. W. LaTourrette

Strombus—top, in allusion to the spiral shell; *gigas*—giant.
Both generic and specific names are of Greek derivation.
One seldom attributes force or speed to gastropods. The
series of photographs shown here attest to these capabilities
as the conch effects a flip in order to right itself, utilizing
the pointed operculum to gain purchase on the sandy
bottom. Cooked in chowder or raw, the species is a
culinary delight.

36

Phylum Mollusca
Class Gastropoda
Order Archaeogastropoda
Family Cypraeidae

ATLANTIC DEER COWRIE
Cypraea cervus Linnaeus
by Don Renn

Cypraea—from the Greek name Kypris for Aphrodite; *cervus*—Latin for deer. The partially extended mantle hides the spots on the base of the shell. These are never ocellated as are those in *C. zebra*. Cowries thrive everywhere in rocky shores to shallow reefs, but are seldom seen during the day. Carefully turn over a number of rocks. You will soon find one or two nestled safely on the underside, along with other interesting creatures. Once the prize is taken, always return the rocks or corals to their original positions!

37

MEASLED COWRIE
Cypraea zebra Linnaeus
by J. W. LaTourrette

Cypraea—a Greek name for Aphrodite; *zebra*—an Abyssinian name for a striped equine of Africa. Cowries have ranked high in favor among collectors, but aquarists should be wary. As is the case with most gastropods, live specimens, when disturbed, emit a perfectly clear, thick discharge that will snuff out all life in the artificial environment.

Phylum Mollusca
Class Gastropoda
Order Archaeogastropoda
Family Ovulidae

FLAMINGO TONGUE
Cyphoma gibbosum (Linnaeus)
by J. W. LaTourrette

Cyphoma—Greek for humpbacked; *gibbosum*—Latin, meaning the same or protuberant. The mottled pattern on the fleshy mantle covering the shell shown here closely resembles the netlike structure of the sea fan over which it browses. The small pink shell is far less attractive than the living specimen. Thus the aquarist gains the edge over the collector and is rewarded with hours of enjoyment and interest in observing the creatures as they live in the sea.

Phylum Mollusca
Class Gastropoda
Order Archaeogastropoda
Family Cassididae

38

SCOTCH BONNET
Phalium granulatum (Born)
by J. W. LaTourrette

Phalium—from the Greek word for white; *granulatum*—Latin for grainy. One of the smaller helmet shells that burrow under the bottom. They can be found while beachcombing during early morning at low tide; frequently little more than a small area of shell betrays their presence beneath the sand.

Phylum Mollusca
Class Gastropoda
Order Archaeogastropoda
Family Cassididae

FLAME HELMET

Cassis flammea (Linnaeus)
by Peter Vila

Cassis—helmet; *flammea*—flaming. Both the generic and specific names are derived from Latin. The parietal shield is generally rounded and it and the outer lip are yellowish cream in color. Spaces between the teeth are the same, without darker areas between.

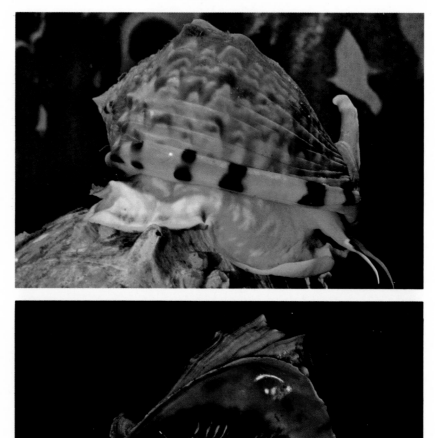

39

EMPEROR HELMET

Cassis madagascariensis Lamarck
by Don Renn

Cassis—helmet in Latin; *madagascariensis*—of Madagascar. Note the black color between the teeth lacking in *C. flammea*. A favorite food of the emperor helmet is the long-spined sea urchin, whose spines it crushes with its massive parietal shield.

Phylum Mollusca
Class Gastropoda
Order Archaeogastropoda
Family Cassididae

CLENCH'S HELMET
Cassis madagascariensis spinella Clench
by Peter Vila

See the preceding caption for the etymology of the generic and specific names; *spinella*—a patronymic subspecies name. This differs from the emperor helmet in that it has smaller, more regular, and far more numerous tubercles on the outer surface of the shell. Both grow to nearly basketball proportions.

40 **Phylum** Mollusca
Class Gastropoda
Order Archaeogastropoda
Family Ficidae

COMMON FIG SHELL
Ficus communis Röding
by J. W. LaTourrette

Ficus—fig in Latin; *communis*—Latin for common. That so large an animal mass can emerge from the fragile home is to me incredible. Beneath the mass is a tan-colored shell so frail that it can be unwittingly crushed between the fingers.

Phylum Mollusca
Class Gastropoda
Order Neogastropoda
Family Muricidae

APPLE MUREX
Murex pomum Gmelin
by Don Renn

Murex—Latin for purple shell; *pomum*—apple in Latin.
The group of specimens shown here have been captured in
the act of laying, fertilizing, and encapsulating eggs in
the ivory-colored leathery pouches. They are hardy
aquarium dwellers; being good browsers, they help keep
the bottom free of organic debris.

41

GIANT EASTERN MUREX
Murex fulvescens Sowerby
by J. W. LaTourrette

Murex—purple shell, Latin; *fulvescens*—also from the
Latin for tawny, dusky, or yellowish. The species bears
longer strong spines than those of *M. pomum*. When
cleaned of surface growths, the shell is white to gray in
color. In the aquarium they have been observed to attack,
kill, and feed on whelks four to five times their size.

Phylum Mollusca
Class Gastropoda
Order Neogastropoda
Family Muricidae

DELTOID ROCK SHELL
Thais deltoidea (Lamarck)
by J. W. LaTourrette

Thais—a celebrated courtesan of Athens; *deltoidea*—Greek for triangular. This small species bears a strong, thick shell. It is found in abundance on intertidal rocks exposed to the ocean surf.

Phylum Mollusca
Class Gastropoda
Order Neogastropoda
Family Melongenidae

42

LIGHTNING WHELK
Busycon contrarium Conrad
by Don Renn

Busycon—a combination of the Anglo-Saxon word *busy* and the Latin word for with; *contrarium*—Latin for against or opposite, in reference to the left-handed direction of the twist of the shell (specimens bearing a right twist are found rarely). The *Busycon* shown here is laying a spiral strand of eggs. Each leathery disk contains numerous eggs, which are easily hatched in the aquarium. Strands are often found on the beach after the juveniles have hatched, each having escaped through the hole always found on the flat surface of every disk. The plentiful whelks serve as a favorite food for the horse conch.

Phylum Mollusca
Class Gastropoda
Order Neogastropoda
Family Fasciolariidae

TRUE TULIP
Fasciolaria tulipa (Linnaeus)
by J. W. LaTourrette

Fasciolaria—striped; *tulipa*—tulip. Both names are derived from Latin. The shell color of this common species may range from orange to black, but always bears the lined pattern. It is easily baited with a cloth bag containing pieces of fish.

The banded tulip, *F. hunteria* Perry, displays characteristic widely spaced, unbroken, purple-brown, spiral bands. Both are carnivores, and if placed in an aquarium with other live shells, will eat the lot. They are common in shallow grassy areas and not infrequently are left behind by the receding tide.

43

HORSE CONCH
Pleuroploca gigantea (Kiener)
by J. W. LaTourrette

Pleuroploca—from two Greek words meaning twisted ribs, an allusion to the general appearance of the shell; *gigantea*—Greek for giant. The chief predator of all conchs, this carnivorous, powerful, and largest of American gastropods has no match among the shellfish. A dark brown periostium covers the shell. When dried, the periostium may flake off, exposing the ivory calcareous material beneath. Younger specimens, to about 3.5 inches in length, display a bright rust-red shell.

Phylum Mollusca
Class Gastropoda
Order Neogastropoda
Family Olividae

LETTERED OLIVE
Oliva sayana Ravenel
by J. W. LaTourrette

Oliva—Latin for olive; *sayana*—a patronymic name. In abundance along Florida's west coast, *Oliva* are easily traced to their locations beneath the sand. Follow furrows across the flats at low tide; dig a few inches at trail's end for the handsomely enameled prize.

Phylum Mollusca
Class Gastropoda
Order Neogastropoda
44 **Family** Marginellidae

Phylum Mollusca
Class Gastropoda
Order Tectibranchia
Family Bullidae

COMMON ATLANTIC MARGINELLA
Prunum apicinum (Menke)
by Peter Vila

Prunum—plum in Latin; *apicinum*—etymology obscure; possibly from the Italian words *apice*, a peak or summit, or *piccino*, little or small. A stroll along shallow flats or shore will often be rewarded with a handful or more of these highly enameled gastropods, each seldom more than one-fourth of an inch in size.

COMMON WEST INDIAN BUBBLE
Bulla occidentalis A. Adams
by J. W. LaTourrette

Bulla—hollow, thin-walled, rounded, bony prominence or vesicle in Latin; *occidentalis*—Latin for western. Collectors should find this common species on grassy mud flats at low tide, especially at night. Empty shells are almost always encountered among debris along the shore. Although frail, they are light in weight and most often will be undamaged.

Phylum Mollusca
Class Gastropoda
Order Ascoglossa
Family Elysiidae

RIBBON NUDIBRANCH*
Tridachia crispata Mörch
by J. W. LaTourrette

Tridachia—etymology obscure; *crispata*—Latin for curling, undulating. Although the color patterns are strikingly dissimilar, all are of the same genus and species. The delightfully inquisitive expression of these shell-less mollusks is as stimulating as their beauty and grace of form. Development of an environmental system for maintaining nudibranchs has eluded my efforts to date. Perhaps the so-called balanced aquarium gaining rapidly in popularity among hobbyists will be the answer. Whatever it is, those who achieve the goal will reap a rich reward.

* Common name suggested by the author.

45

Phylum Mollusca
Class Gastropoda
Order Anaspidea
Family Aplysiidae

SPOTTED SEA HARE
Aplysia dactylomela Rang
by Don Renn

Aplysia—from the Greek for unwashed or filthy, perhaps in allusion to its soft, slimy texture; *dactylomela*—Greek words meaning black finger. The shell of this mollusk is internal, very thin, and contains little or no lime; slight pressure with the finger on the posterior dorsal surface will reveal its presence within. Disturbed, sea hares discharge a staining but harmless purple ink.

46

BLACK SEA HARE
Aplysia morio (A. E. Varrill)
by J. W. LaTourrette

Aplysia—unwashed in Greek; *morio*—related to the Greek *moros*, stupid. This strong and graceful swimmer attains a length of 1 foot or more. Apparently distasteful, it enjoys a life relatively free from predation. *Aplysia* can be maintained in open-system aquaria if supplied with ample algae on which to feed.

Phylum Mollusca
Class Gastropoda
Order Anaspidea
Family Aplysiidae

GREEN SEA HARE*
Dolabrifera dolabrifera Rang
by Peter Vila

Dolabrifera—both the generic and specific names are Latin for hatchet-shaped. The green color and rough texture camouflage this creature as it browses among algal growths. The bright green sole is flattened on the aquarium glass in the ventral view. Unlike the shells of the other two sea hares depicted in this book, the internal shell of *Dolabrifera* is heavily calcified and covered by a brown periostracum.

* Common name suggested by the author.

47

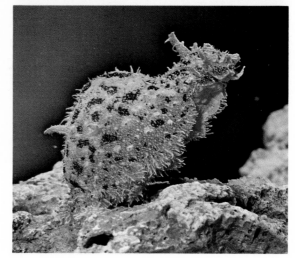

RAGGED SEA HARE
Bursatella leachi plei Rang
by J. W. LaTourrette

Bursatella—Latin for little purse or bag; *leachi plei*—patronyms. Unlike *Aplysia* species, adult ragged sea hares contain no vestigial molluscan shell. The common name reference to the hare is obvious: long fleshy anterior sensory projections resemble rabbit ears.

Phylum Mollusca
Class Gastropoda
Order Notaspidea
Family Pleurobranchidae

NUBBY SEA SLUG*
Pleurobranchus areolatus Mörch
by J. W. LaTourrette

Pleurobranchus—Greek for side gill; *areolatus*—with small spaces, Latin. Intriguing in knobby texture and color pattern, when placed in an invertebrate aquarium this specimen further piqued my interest by laying ribbony coils of eggs. A whelk (*Busycon contrarium*) found them a tasty treat and robbed me of the prize.

* Common name suggested by the author.

48 **Phylum** Mollusca
Class Gastropoda
Order Notaspidea
Family Polyceridae

 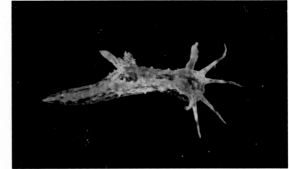

HORNED NUDIBRANCH*
Polycera hummi Abbott
by Peter Vila

Polycera—Greek and Latin words for many horned or tendriled; *hummi*—a patronymic name. Even the adult of this species is so small that it is capable of creeping along in an upside-down position on the water's surface film. Bland to the naked eye, *Polycera* displays true colors under magnification and photographic light.

* Common name suggested by the author.

Phylum Mollusca
Class Gastropoda
Order Doridoidea
Family Dorididae

SEA CAT NUDIBRANCH*
Felimare bayeri Marcus
by J. W. LaTourrette

Felimare—Latin for sea cat; *bayeri*—a patronymic name. To date the species is known only from four specimens: one taken in 1963, a second in 1965, and the two shown here, which were found with their egg masses on a sheet of metal by the Seaquarium in 1967. All were found in Biscayne Bay. Egg masses and adults were taken to the University of Miami Rosenstiel School of Marine and Atmospheric Science. Observations of hatching and early-stage development afforded a wealth of data concerning this rare molluscan form.

* Common name suggested by the author.

Phylum Mollusca
Class Gastropoda
Order Doridoidea
Family Dorididae

GREELEY'S NUDIBRANCH*
Peltodoris greeleyi Marcus & Marcus
by J. W. LaTourrette

Peltodoris—etymology obscure; possibly from an old French word, *pel*, for the skin of a fur-bearing animal. Doris was a Grecian sea goddess. The specific name, *greeleyi*, is a patronym. The first part of the generic name may refer to the rather fuzzy dorsal epidermis. In spite of gross differences in pattern and color, the specimens are of the same species. Note that the larger orange one has withdrawn its delicate naked gills (nudibranchs). Rhinophores are retractable as well, a protective device for these delicate vital organs.

* Common name suggested by the author.

50

LEATHERY NUDIBRANCH*
Platydoris angustipes (Mörch)
by J. W. LaTourrette

Platydoris—from the Greek words for flat and sea goddess (also a word for a leather sac); *angustipes*—narrow in Latin. Although soft in appearance, the depressed body feels like stiff leather to the touch. The corona of gills (circumanal tuft) surrounds the anus in family members.

* Common name suggested by the author.

Phylum Mollusca
Class Gastropoda
Order Eolidoidea
Family Fionidae

FEATHERED NUDIBRANCH*
Fiona pinnata (Eschscholtz)
by J. W. LaTourrette

Fiona—a semimythical race of Gaelic warriors of supernatural size, strength, and daring; *pinnata*—Latin for feathered. The specimens shown here are brown from feeding on gooseneck barnacles affixed to the piece of wood. When associated with, and feeding on *Velella*, they are purple in color. Neat spiral egg masses are attached to the wood by a strong single strand. Larval *Fiona* hatch in two days and reach maturity in two weeks.

* Common name suggested by the author.

Phylum Mollusca
Class Gastropoda
Order Eolidoidea
Family Facelinidae

51

ORANGE-LINED NUDIBRANCH
Learchis poica Marcus & Marcus
by J. W. LaTourrette

Learchis—etymology obscure; *poica*—possibly a short form of a Greek word for many-colored. This common nudibranch bears a distinctive orange line along either side of the body and another between the rhinophores on the dorsal surface of the head. To the left it is seen traversing the apex of a channeled whelk (*Busycon canaliculatum*) to the aquarium glass.

Phylum Mollusca
Class Gastropoda
Order Eolidoidea
Family Aeolidiidae

SPURRED NUDIBRANCH*
Spurilla neapolitana (Delle Chiaje)
by Peter Vila

Spurilla—from the Anglo-Saxon word *spur* and the Latin diminutive suffix for little (i.e., covered with little spurs); *neapolitana*—of Naples. Circular white spots on body and cerata, and the orange tinge to epidermis and viscera are distinctive. This hardy little mollusk feeds on anemones.

* Common name suggested by the author.

Phylum Mollusca
Class Gastropoda
Order Eolidoidea
Family Glaucidae

52

BLUE GLAUCUS
Glaucus marinus DuPont
by Don Renn

Glaucus—Latin for bluish-green or sea-colored; *marinus*—Latin for marine. These pelagic nudibranchs float upside down at the water's surface, a practical attitude enabling them to feed on the tentacles of *Velella, Porpita,* and *Physalia.* Stinging nematocysts of the latter—undigested, untriggered, and stored in feathery cerata—serve admirably as defensive ammunition. As larvae, these mollusks are protected within a tiny coiled shell that is lost with age.

Phylum Mollusca
Class Gastropoda
Order Eolidoidea
Family Scyllaeidae

SARGASSUM NUDIBRANCH (or SEA SLUG)
Scyllaea pelagica Linné
by Peter Vila

Scyllaea—from Skylla, a mythological Greek sea monster with twelve arms and six necks; *pelagica*—of the sea, Greek. Leaflike rhinophores and gills, and an adaptive color pattern make this small creature almost impossible to see in the sargassum weed environment. A firm shake of a clump of the weed usually will yield one or two, as well as a host of small fishes, crustaceans, and other interesting forms of life.

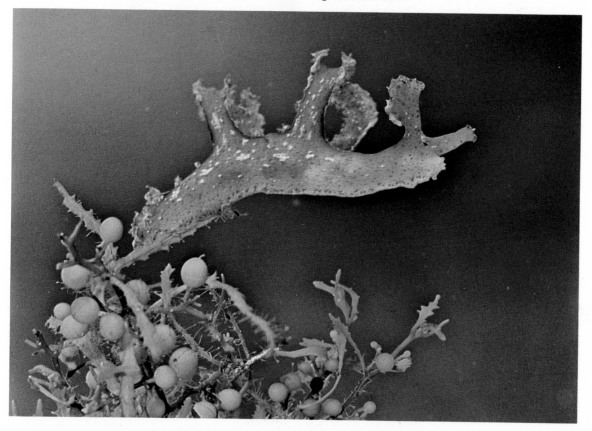

53

Phylum Mollusca
Class Amphineura
Order Chitonida
Family Chitonidae

FUZZY CHITON
Acanthopleura granulata (Gmelin)
by Don Renn

Acanthopleura—thorny side in Greek; *granulata*—granular in Latin. The shell of the chiton is composed of eight hinged plates. At night the animals range forth to feed; morning finds each returned to its original niche in the rock.

Phylum Mollusca
Class Pelecypoda
Order Filibranchia
Family Pectinidae

54

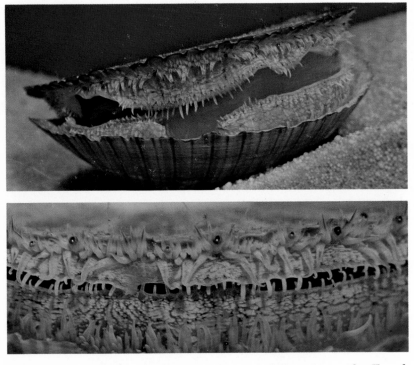

ZIGZAG SCALLOP
Pecten ziczac (Linnaeus)
by J. W. LaTourrette

Pecten—Latin for comb or comblike; *ziczac*—the French word for zigzag, alternate change of direction by sharp angles. The specific name was derived from the mode of propulsion: snapping of the valves moves the scallop in a rapid jerky fashion. A subject of intensive mariculture study, scallops return investigative scrutiny with a multitude of bright blue eyes.

Phylum Mollusca
Class Pelecypoda
Order Filibranchia
Family Spondylidae

ATLANTIC THORNY OYSTER
Spondylus americanus Herman
by Peter Vila

Spondylus—vertebrata in Latin, perhaps alluding to spiny processes on the shell; the specific name is obvious. Surface growths such as the sponge shown here and the fire coral (see *Millepora alcicornis*) often mask the malacologist's prize beneath. These are often found clingling to wrecks and, taken with care, are hardy in open-system aquaria. In the sea they feed on plankton; in the aquarium I substitute brine shrimp, fish blood, and 10% dextrose solution.

55

Phylum Mollusca
Class Pelecypoda
Order Filibranchia
Family Limidae

DELICATE LIMA*
Lima scabra tenera Sowerby
by J. W. LaTourrette

Lima—Latin for file; *scabra*—rough; *tenera*—delicate. Both the specific and subspecific names are derived from Latin. A more finely ribbed and more delicate shell than that found in the rough lima cause this file shell to be classified as a subspecies.

* Common name suggested by the author.

Phylum Mollusca
Class Pelecypoda
Order Filibranchia
Family Limidae

ROUGH LIMA
Lima scabra (Born)
by J. W. LaTourrette

Lima—file; *scabra*—rough. The generic and specific names are derived from Latin. Brilliant hues and a reasonable hardiness give this mollusk top rank among aquarists. Although different in color, the two specimens shown here are of the same species. In the second type of this organism, the delicate lima, the shell is smoother, lacking the filelike ridges that are referred to in both the generic and family names.

56

Phylum Mollusca
Class Pelecypoda
Order Filibranchia
Family Pholadidae

WEDGE-SHAPED MARTESIA
Martesia cuneiformis (Say)
by J. W. LaTourrette

Martesia—from Greek or Latin meaning of the
sea; *cuneiformis*—Latin for wedge-shaped. An
auger-shaped shell enables this mollusk to bore
through submerged wood efficiently. The shape
of the shell is widely divergent among individuals,
probably from encountering areas of varying
density within the wood. A more delicate relative
of *Martesia* burrows in soft mud: *Barnea costata*,
the lovely angel wing.

Phylum Mollusca
Class Pelecypoda
Order Filibranchia
Family Gastrochaenidae

SPENGLER CLAM
Spengleria rostrata Spengler
by Peter Vila

Spengleria—a patronym; *rostrata*—pointed in
Latin. Small holes on the underside of corals
reveal the presence of this borer, seen here removed from the safety of its stony cave.

Phylum Mollusca
Class Cephalopoda
Order Decapoda
Family Loliginidae

ATLANTIC OVAL SQUID
Sepioteuthis sepioidea (Blainville)
by Mike Davis

Sepioteuthis—Latin and Greek terms for cuttlefish or squid ink; *sepioidea*—from the Greek word *sepion*, the internal shell of the squid. In the species shown here, this is reduced to a long, translucent, dart-shaped form. Few sights in the marine world equal that of the ballet of the squid. Intelligent and seemingly perfectly adapted to their environment, they elude our efforts and fare poorly in the aquarium.

58 **Phylum** Mollusca
Class Cephalopoda
Order Decapoda
Family Ommastrephidae

ORANGE BACK SQUID
Ommastrephes pteropus Steenstrup
by Don Renn

Ommastrephes—eye and flash of lightning; *pteropus*—winged. Both names are derived from Greek. It is larger than the Atlantic oval squid and a stronger swimmer, bearing large posteriorly attached fins (the wings alluded to in the specific name).

Phylum Mollusca
Class Cephalopoda
Order Octopoda
Family Argonautidae

PAPER NAUTILUS*
Argonauta argo Linnaeus
Live specimens by J. W. LaTourrette
Shell with egg mass by Don Renn

Argonauta—a combination of two Greek words meaning white sailor; *argo*—white in Greek. The female specimen in the photograph to the left was taken alive in Biscayne Bay and survived for several days in an aquarium. In the upper photograph the mantles borne by the first arms are withdrawn, exposing the frail porcelaneous shell. In the lower photograph the sensitive mantle is extended, clearly displaying the chromatophores, pigment cells that expand and contract to reveal purple through coppery hues. The protruding siphon passes water over the gills and serves as an efficient jet-propulsion unit. During her brief stay she consumed small live fishes placed in the beak located in front of the eyes at the apex of the four pairs of arms.

The shell shown here bears an egg mass and five hectocotyli, sperm-bearing autonomous third left arms of the male of the species. Males are very small in comparison to females. Their hectocotylized arm is usually carried coiled up in a sac beneath the left eye. Thus the alluring lady that had formed the egg case had no less than five suitors. Two hectocotyli appear as wormlike individuals on the aquarium glass; the remaining three are among the eggs. Stimulated by heat from the photographic lights, a hectocotylus delivers its threadlike sperm mass in the inset above.

* Some malacologists prefer the common name "argonaut shell" because this cephalopod is not a nautilus, but only bears superficial resemblance to the chambered nautilus, *Nautilus pompilius*. The specific name *argo* may also refer to the name of the ship in which Jason sailed in search of the golden fleece.

Phylum Mollusca
Class Cephalopoda
Order Octopoda
Family Octopodidae

BRIAR OCTOPUS
Octopus briareus Robson
by J. W. LaTourrette

Octopus—eight in Latin and foot in Greek; *briareus*—from the Anglo-Saxon word briar or thorn. It is smaller than O. *vulgaris*, with long, thin arms between which stretches a delicate web. Octopods in captivity breed freely and often glue long strings of many clear eggs on the glass, enabling their embryonic development to be observed.

61

DWARF OCTOPUS
Octopus joubini Robson
by Don Renn

Octopus—eight-footed (armed) in Latin and Greek; *joubini*—a patronymic name. The tiny *joubini* is the enigma of octopods within its range. It is the smallest of the species included here, but lays the largest eggs, possesses the most virulent bite, and does not hesitate to utilize this defense. Nine times out of ten they are brought in by collectors who were working in shallow water and found them in clam shells.

Phylum Mollusca
Class Cephalopoda
Order Octopoda
Family Octopodidae

62

WHITE SPOTTED OCTOPUS
Octopus macropus Risso
by J. W. LaTourrette

Octopus—eight-footed in Latin and Greek; *macropus*—Greek for long foot, as manifested by the length of the arms in the photographs. Swimming (top) and walking (bottom) attitudes are illustrated here. The brown haze in the water is sepia expelled by the octopus as a mechanism of defense. This is harmless diluted in the sea; in an aquarium with a recirculating water system, unless the water is exchanged at once, everything therein will die. This specimen inked during an attack on the crab, a favorite food of octopods. The crab, *Eriphia gonagra*, pinched an arm with its stout claw and eluded its antagonist.

Phylum Mollusca
Class Cephalopoda
Order Octopoda
Family Octopodidae

63

COMMON ATLANTIC OCTOPUS
Octopus vulgaris Lamarck
by J. W. LaTourrette

Octopus—Latin and Greek words are combined to mean eight-footed; *vulgaris*—common in Latin. The two photographs shown here are of the same specimen, taken minutes apart. They illustrate perfectly the ability of these cephalopods to change color as well as skin texture. Long overrated as one of the horrors of the seas, in reality they are timid.

PHYLUM ANNELIDA
THE SEGMENTED WORMS

The classes Polychaeta, the paddle-footed annelids; Oligochaeta, the bristle-footed annelids; and Hirudinea, the leaches, are contained within the phylum Annelida. More than 6000 species are known. They range in diverse habitats from high glacial environments to beneath the earth's soil, through fresh water to the depths of the sea. The sea provides the habitat apparently most suitable to the greatest number of species. In turn, annelids serve as a food staple for an immense variety of creatures. This renders them of considerable commercial value. The rather repugnant leaches provide medicine with basic blood anticoagulants. In less enlightened times they served as living instruments for bleeding and removing surface blood clots, as in the case of a black eye caused by colliding with a door knob or some other swiftly moving solid object. Certain other forms of segmented worms are utilized as food for human consumption. Others, whether real or artificial, serve the fishing-bait industry. Still others—alive, dried, or frozen—are purchased in quantities amounting to hundreds of thousands of dollars annually by the aquarium trade.

Three marine forms have been photographed and are included as representatives of this phylum. Note that the serpulid worms are unidentified as to genus and species. In order to study these systematically, they must be removed from their calcareous tubes. They continue to flourish within their display aquarium, so we must be satisfied with a family name alone.

Phylum Annelida
Class Polychaeta
Order Sabellida
Family Sabellidae

BLACK-SPOTTED SABELLID WORM*
Sabella melanostigma Schmarada
by Peter Vila

Sabella—sand in Greek; *melanostigma*—Greek for black spot, alluding to paired eye spots on the tentacles of the crown. The parchmentlike tubes of a multitude of these worms are affixed along with algae and other growths to the substrate.

* Common name suggested by the author.

Phylum Annelida
Class Polychaeta
Order Sabellida
Family Serpulidae

65

SERPULID WORM
Genus and species unknown
by Peter Vila

The word "serpulid" is derived from the Latin *serpula*, meaning little snake. The two color forms shown here may not be of the same species. Identification is impossible without dissecting the animal from its calcareous tube within the branches of coral (*Oculina diffusa*). The specimens continue to thrive, so I must be content with the family name alone. Both display their tentacular crown in the feeding position.

Phylum Annelida
Class Polychaeta
Order Amphinomida
Family Amphinomidae

66

BRISTLE WORM
Chloeia viridis Schmarada
by Mike Davis and Peter Vila

Chloeia—Greek, in reference to young shoots of grass; *viridis*—green in Latin, from the color of the holotype specimen. If handled, fine, hairlike spines penetrate the skin, break off, and result in a painful wound and infection. Under ultraviolet light the spines glow emerald green. Bristle worms are found among bottom debris and under floating objects, such as life rafts. They are listed in the U.S. Air Force *Survival at Sea* manual as a potential hazard.

PHYLUM ARTHROPODA
THE ARTHROPODS

The insects are assumed by most people to live in terrestrial habitats. With but a handful of exceptions this would seem to be the general rule: there are spiders that lead a truly aquatic existence and one that breathes air by transporting bubbles to its personal diving bell beneath some tropical Pacific coral reef. However, exclusive of insects, the large and fantastically diverse phylum Arthropoda boasts at least an additional 65,000 species members. Of these, two classes are certain to be encountered throughout Florida and Bahamian waters: Arachnomorpha and Crustacea. The former, more closely allied to the spiders and spiderlike insects than its vernacular name would lead one to believe, is *Limulus*, the horse-shoe crab or horse-foot crab as it was more aptly called in the late nineteenth century. The latter, *Crustacea*, encompasses a multitude of creatures that one can hardly avoid even several miles from the sea (land crabs, land hermit crabs, and the like). In fact, unless polluted to total decimation, no beach or marine area will be without a handsome cross section of crustaceans. Included in this class are the multitude of barnacles on rocks, shore, other living animals, and floating debris. Shrimps, lobsters, and crabs surpassing the wildest imagination in appearance are liable to be encountered almost anywhere. Beyond their value as interesting life forms is the very real commercial importance of crustaceans. Blue and stone crabs as well as several species of lobster and shrimp are harvested in abundance. Maricultural efforts to raise these gourmet delights have been under way for a number of years in order to meet the increasing demands on the fishing industry. Many other species of Crustacea are edible—for example, the mantis shrimps, but are not available in commercial quantities. Perhaps mariculturists will see fit to add these, locust lobsters, or even some of the very large edible barnacles to their lists for investigation. Do not misinterpret this to mean that all crustaceans are edible. There are some that are highly toxic; members of the family Calappidae are known to be extremely dangerous and came close to claiming the life of a personal acquaintance. Thus, if one feels compelled to experiment in this vein, I urge prior reference to the voluminous work by Halstead listed in the bibliography.

Phylum Arthropoda
Class Arachnomorpha
Order Merostomata
Family Limulidae

HORSESHOE CRAB
Limulus polyphemus Linnaeus
by Mike Davis

Limulus—from the Latin *limus*, sidelong; *polyphemus*—Polyphemus, the many-voiced giant of Greek mythology. This is not a true crab, but a member of the class Arachnomorpha, which includes the scorpions and spiderlike arthropods. *Limulus* today displays little if any change from fossil remains. Heavy armor plating studded with sharp spines protects the creature in one of Nature's finer displays of environmental adaptation. Hardy young specimens are interesting subjects for home aquarists and are found with ease in shallow water along the entire length of the American coast to the Yucatan.

Phylum Arthropoda
Class Crustacea
Order Cirripedia
Family Balanidae

BARNACLE
Chelonibia patula (Ranzani)
by J. W. LaTourrette

Chelonibia—Greek for cloven, as in a hoof or claw, and water bucket; *patula*—Latin for expanding. The barnacle in this photograph is firmly cemented to the shell of a live crab. Cursed for centuries for fouling ships' hulls and rasping the flesh of swimmers, the creatures now are studied in an effort to learn the secret of their cementing ability.

Phylum Arthropoda
Class Crustacea
Order Cirripedia
Family Lepadidae

GOOSENECK BARNACLE
Lepas ansifera Linnaeus
by J. W. LaTourrette

Lepas—Greek for limpet; *ansifera*—gooselike in Latin. Both *Lepas* and *Chelonibia* have been captured in these pictures in the act of seining planktonic food from the water with their frail skein of extended feet. Boaters and beachcombers most often will find gooseneck barnacles attached to all manner of floating debris. In some parts of the world larger species are utilized for human consumption. Certain nudibranchs (see FEATHERED NUDIBRANCH) and other animals find them tasty as well.

Phylum Arthropoda
Class Crustacea
Order Stomatopoda
Family Lysiosquillidae

MANTIS SHRIMP
Lysiosquilla scabricauda (Lamarck)
by Mike Davis

Lysiosquilla—Greek for releasing or loosening and Latin for a prawn or shrimp; *scabricauda*—a combination of two Latin words meaning rough tail. Comparable to the insect praying mantis in both appetite and large claws. Fishermen call them "thumbcracker," a name of obvious meaning if you try to handle a live specimen. Periscopelike eyes protrude from burrows in soft sand or mud, with the tail, which can be up to a foot in length, remaining well concealed. Although edible, the species is not found in commercial quantities.

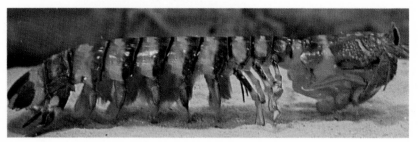

Phylum Arthropoda
Class Crustacea
Order Stomatopoda
Family Gonodactylidae

FALSE MANTIS SHRIMP*
Pseudosquilla ciliata (Fabricius)
by J. W. LaTourrette

Pseudosquilla—Greek for false and Latin for shrimp; *ciliata*—a Latin word meaning furnished with cilia, hairlike processes. The forms, shown here are of the same genus and species: another fine example of the fact that color serves little function in systematics.

* Common name suggested by the author.

Phylum Arthropoda
Class Crustacea
Order Stomatopoda
Family Squillidae

ARMORED MANTIS SHRIMP*
Squilla rugosa Bigelow
by J. W. LaTourrette

Squilla—prawn or shrimp in Latin; *rugosa*—Latin for wrinkled. Exoskeleton ridges define geometric plates on this interesting form. I suggest the aquarist think twice before placing *Squilla* in the same tank with valuable small fish; they are adept at snaring live fish, and in no time some prized specimens will be among the missing.

* Common name suggested by the author.

Phylum Arthropoda
Class Crustacea
Order Decapoda
Family Penaeidae

70

PINK SHRIMP
Penaeus duorarum Burkenroad
by J. W. LaTourrette

Penaeus—the legendary father of Daphne of Greek myth; *duorarum*—from Latin words meaning two and rostrum or nose. This is the species best known as a table delicacy. They spawn in south Florida estuaries and migrate to more saline waters as they grow. Intense studies in artificial spawning and rearing will undoubtedly result in successful mariculture operations on a commercial scale. Most people will enjoy them even more if they catch them themselves. Little is required in the way of special equipment: a long-handled net and lantern will suffice. Find a spot among the crowd of people on a bridge catwalk, attract the shrimp with the lantern, and net them without falling in.

Phylum Arthropoda
Class Crustacea
Order Decapoda
Family Penaeidae

ROCK SHRIMP
Sicyonia brevirostris Stimpson
by J. W. LaTourrette

Sicyonia—a mythological name; also Greek for wild cucumber or gourd, perhaps in allusion to its rough surface texture; *brevirostris*—Latin for short-snouted. Fine table fare, the value of the species is limited only by rather small commercial catches. Future studies of life cycle, spawning habits, food sources, and the like might well spotlight this hardy aquarium shrimp as a candidate for marine farming. Limited commercial availability could reward the successful mariculturist with premium returns.

PARR'S ROCK SHRIMP*
Sicyonia parri (Burkenroad)
by J. W. LaTourrette

Sicyonia—a mythological name; *parri*—a patronym. Found while seining in a shallow bed of eel grass, the specimen displayed a protective green color phase. Placed in a bland aquarium, it promptly assumed a tan hue to match the sand under which it burrowed. Returned to a green plastic pail, the shrimp again assumed the darker green coloration.

* Common name suggested by the author.

Phylum Arthropoda
Class Crustacea
Order Decapoda
Family Palaemonidae

PETERSON SHRIMP
Periclimenes petersoni Chace
by J. W. LaTourrette

Periclimenes—name assigned by Hesiod to a son of Poseidon; *petersoni*—an honorary name. A small species that lives commensally with sea anemoes. The crystal clarity of the exoskeleton reveals its presence only when reflecting silverlings of light; royal-purple markings verify their being.

72

SPOTTED CLEANER SHRIMP
Periclimenes yucatanicus (Ives)
by J. W. LaTourrette

Periclimenes—a name assigned by Hesiod to a son of Poseidon; *yucatanicus*—of Yucatan, where the species was probably first taken. Another small commensal symbiont that associates mostly with the anemones *Condylactis gigantea* and *Bartholomea annulata*.

Phylum Arthropoda
Class Crustacea
Order Decapoda
Family Palaemonidae

TWO CLAWED SHRIMP*
Brachycarpus biunguiculatus (Lucas)
by J. W. LaTourrette

Brachycarpus—Greek and Latin words for short wrist; *biunguiculatus*—two-clawed in Latin. This is at least one species of decapod crustacean that is virtually pantropical in geographic distribution.

* Common name suggested by the author.

73

PEN SHRIMP*
Pontonia mexicana Guerin
by Peter Vila

Pontonia—Latin for a small boat or pontoon (the name also may stem from the Spanish *pontos*, meaning of the sea); *mexicana*—Mexican. These small shrimps live in a commensal relationship within the shell of *Pinna* species. The body of the female (upper right) appears more colorful than the male due to the egg mass she carries.

* Common name suggested by the author.

Phylum Arthropoda
Class Crustacea
Order Decapoda
Family Stenopidae

LIMA SHRIMP*
Microprosthema semilaevis (von Martens)
by J. W. LaTourrette

Microprosthema—derived from two Greek words meaning small appendages; *semilaevis*—Latin for half and small, light, or nimble. This strikingly beautiful specimen, three-fourths of an inch long, was taken in association with a *Lima scabra*. When placed in an aquarium with other *Lima*, it stayed near their scarlet tentacles, but was not observed to enter the shell.

* Common name suggested by the author.

74

BANDED CORAL SHRIMP
Stenopus hispidus (Olivier)
by Don Renn

Stenopus—Greek for narrow or straight; *hispidus*—spiny, shaggy, or rough in Latin. A long-time aquarium favorite, this delicate species serves an important function both in nature and the artificial environment. Fishes seek them out in order to be cleansed of parasites and dead tissue. So popular is *Stenopus* that mariculturists are attempting to breed them solely for the aquarium field.

Phylum Arthropoda
Class Crustacea
Order Decapoda
Family Stenopidae

GOLDEN CORAL SHRIMP
Stenopus scutellatus Rankin
by J. W. LaTourrette

Stenopus—narrow or straight, Greek; *scutellatus*—Latin, meaning armed with a shield or shielded (by body plates). This species is easily distinguished from *S. hispidus* due to its golden color. The female shown here bears a greenish egg mass. She has lost her major claw, but will regenerate a replacement in time.

75

Phylum Arthropoda
Class Crustacea
Order Decapoda
Family Gnathophyllidae

BUMBLE BEE SHRIMP
Gnathophyllum americanum Guerin
by Peter Vila

Gnathophyllum—from two Greek words meaning leafy jaw; the specific name is self-explanatory. The tube feet of sea urchins are considered a delicacy by this small interesting type of shrimp Collectors can find them with some difficulty on grassy bottom areas inhabited by urchins.

Phylum Arthropoda
Class Crustacea
Order Decapoda
Family Alpheidae

BANDED SNAPPING SHRIMP
Alpheus armillatus H. Milne Edwards
by J. W. LaTourrette

Alpheus—Greek for white spots on the skin (exoskeleton); *armillatus*—Latin, consisting of rings. These are found in great numbers among shoreline rocks and on reefs. The specimen shown here was taken with many others along with *Synalpheus brevicarpus* from the passages of a large beached sponge.

76

STRIPED SNAPPING SHRIMP
Alpheus formosus Gibbes
by J. W. LaTourrette

Alpheus—white spots on the skin in Greek; *formosus*—beautiful in Latin. The single large claw of *Alpheus* species produces an audible click; large numbers create a cacophony of sound. Within the confines of an aquarium these shrimp are more often heard than seen.

SNAPPING SHRIMP
Synalpheus brevicarpus Coutiere
by J. W. LaTourrette

Synalpheus—Greek for like *Alpheus*; *brevicarpus*—Latin for short hand or wrist. A green-bodied variation of the species is common. In younger specimens the red claw more closely matches the body color. *Synalpheus brevicarpus* is confused easily with *S. minus*; both are found in dead corals or throughout the passages of sponges.

Phylum Arthropoda
Class Crustacea
Order Decapoda
Family Hippolytidae

SCARLET LADY*
Hippolysmata grabhami Gordon
by Don Renn

Hippolysmata—in reference to Hippolyte, the Queen of the Amazons in Greek myth; *grabhami*—a patronymic name. The female shown here bears a greenish mass of eggs. The species is also given the common name "red-backed cleaner shrimp," alluding to their mutualistic habits.

* Common name suggested by the author.

77

VEINED SHRIMP*
Hippolysmata wurdemanni (Gibbes)
by Don Renn

Hippolysmata—Hippolyte, Queen of the Amazons in Greek myth; *wurdemanni*—a patronym. The color pattern is distinctive of this symbiont, here shown prancing on a bleached stand of *Acropora cervicornis*. The red and white color pattern, particularly when in bands on antennae, may have a correlation with symbiotic behavioral patterns displayed by shrimp species.

* Common name suggested by the author.

Phylum Arthropoda
Class Crustacea
Order Decapoda
Family Homaridae

AMERICAN LOBSTER
Homarus americanus H. Milne Edwards
by Mike Davis

Homarus—from an old French word for lobster; *americanus*—American. Note: this species, an American table delicacy for over three centuries, does *not* live within the geographic area defined for this work! North Carolina is about the southern limit of its range. Comparison with other eastern Atlantic species shows it to be the only one to bear the massive claws. The American, or "Maine," lobster is included here to demonstrate this gross difference.

78

Phylum Arthropoda
Class Crustacea
Order Decapoda
Family Palinuridae

SPINY LOBSTER
Panulirus argus (Latreille)
Adult by J. W. LaTourrette; juvenile by Don Renn

Panulirus—an anagram of *Palinurus*, the Latin name of an arthropod genus; *argus*—Greek, meaning shiny or bright. If the American, or "Maine," lobster is the gourmet delicacy southward to the Carolinas, the spiny, or "Florida," lobster takes over from there. Both are among the few ocean species protected by law (laws vary from year to year). Florida's closed season is from March 31 to July 31. Thereafter they may be taken without use of spears, hooks, or grabs. The carapace must be over 3 inches and the tail no less than $5\frac{3}{4}$ inches. They are to remain intact at all times, and no egg-bearing females can be kept.

ROCK LOBSTER
Panulirus guttatus (Latreille)
by Mike Davis

Panulirus—anagram of the name of an arthropod genus *Palinurus*; *guttatus*—also from a Latin word for spotted. The white-spotted color pattern is distinctive. The species is not found in sufficient abundance to be commercially valuable in its own right; they are included in the spiny lobster catch. Rock lobsters are nocturnal and rarely venture from their cover during daylight. Unlike the spiny lobster, which will eat just about anything, *P. guttatus* fares best in captivity on a molluscan diet.

Phylum Arthropoda
Class Crustacea
Order Decapoda
Family Palinuridae

BRAZILIAN LOBSTER
Panulirus laevicauda (Latrielle)
by J. W. LaTourrette

Panulirus—anagram of *Palinurus*; *laevicauda*—two Latin words for smooth tail. Easily differentiated from the two preceding forms in that white spots are confined to the margins of the tail and the posterior area of the carapace has a definite green cast. Its commercial status is that of *P. guttatus*; its habits and feeding patterns are much the same.

COPPER LOBSTER*
Palinurellus gundlachi von Martens
by Peter Vila

Palinurellus—from the Latin *Palinurus*, pilot of Aneas (aneas—a brazen of copper); *gundlachi*—a patronymic name. A deep-water inhabitant more rare than *Justitia*. The elongate granulated carapace is distinctive.

* Common name suggested by the author.

Phylum Arthropoda
Class Crustacea
Order Decapoda
Family Palinuridae

LONG-ARMED LOBSTER*
Justitia longimanus (H. Milne Edwards)
by J. W. LaTourrette

Justitia—Latin for upright; *longimanus*—Latin for long hand. Not considered rare, but a seldom seen deep-water inhabitant. This species is unique in this range in that it has a fairly large claw on one of the first walking legs.

* Common name suggested by the author.

81

Phylum Arthropoda
Class Crustacea
Order Decapoda
Family Scyllaridae

SPANISH LOBSTER
Scyllarus americanus (Smith)
by J. W. LaTourrette

Scyllarus—a kind of crab, Greek; *americanus*—American. A smaller form than the locust lobster, it is found in a variety of color patterns. This specimen was bearing an egg mass when taken in February.

Phylum Arthropoda
Class Crustacea
Order Decapoda
Family Scyllaridae

LOCUST LOBSTER
Scyllarides aequinoctalis (Lund)
by J. W. LaTourrette

Scyllarides—a Greek term for a kind of crab; *aequinoctalis*—from two Latin terms for of the sea and night or noctural. An oddity in appearance, the locust, or shovel-nosed, lobster is edible, but not found in commercial quantities. Family members lack long antennae; theirs are paddle-shaped and may be used for digging.

Phylum Arthropoda
Class Crustacea
Order Decapoda
Family Axiidae

BURROWING SHRIMP*
Axiopsis sp.
by J. W. LaTourrette

Axiopsis—from Greek terms for of like value or likeness; species unknown. Little is known of the life history of this unique family of shrimps. They dwell in bottom burrows into which grasses and debris are carried for food.

* Common name suggested by the author.

Phylum Arthropoda
Class Crustacea
Order Decapoda
Family Porcellanidae

SPOTTED PORCELAIN CRAB
Porcellana sayana (Leach)
by J. W. LaTourrette

Porcellana—Italian for the nacre of the Venus shell; *sayana*—a patronym. This small species is often found sharing the shell of the hermit crab, *Petrochirus diogenes*. In this case three porcellanids were in a *Busycon* shell; as many as five have been known to share a single habitat.

83

Phylum Arthropoda
Class Crustacea
Order Decapoda
Family Coenobitidae

LAND HERMIT CRAB
Coenobita clypeatus (Herbst)
by J. W. LaTourrette

Coenobita—one of a religious order living in a convent or in a community, in opposition to a hermit, who lives in solitude; *clypeatus*—Latin, meaning provided with a shield. Moisture carried in the gills enables this crustacean to lead a terrestrial life.

Phylum Arthropoda
Class Crustacea
Order Decapoda
Family Paguridae

RED HERMIT CRAB*
Paguristes cadenati Forest
by J. W. LaTourrette

Paguristes—related to *pagouros*, the Greek word for a kind of crab; *cadenati*—a patronym. Members of this genus range into deeper waters, the specimen shown here having been taken from an offshore reef.

* Common name suggested by the author.

84

STRIPED HERMIT CRAB
Clibanarius vittatus (Bosc)
by J. W. LaTourrette

Clibanarius—a Greek word for pot or earthenware vessel and a Latin term for belonging to allude to the adopted shell home; *vittatus*—striped in Latin. Although found on dry rocky areas along the shore, they are unable to extract oxygen from the air and must return to tide pools or sea for moisture.

Phylum Arthropoda
Class Crustacea
Order Decapoda
Family Paguridae

HERMIT CRAB
Petrochirus diogenes (Linnaeus)
by J. W. LaTourrette

Petrochirus—a combination of terms for like a stone and hand or one who moves the hands with regularity as in pantomime; *diogenes*—the Greek philosopher who lived in a cask. The generic name, from the Greek, illustrates the constant motion of small feeding appendages between the large claws. This is the largest of hermit crab species in the Florida area. Gastropod shells protect the soft armorless abdomen. When the crab outgrows its home, the shell is discarded for a larger one. Small crabs (see *Porcellana sayana*) often share the same shell in a fine example of commensalism. The whelk shell at the top is overgrown with barnacles and slipper shells.

85

BAR-EYED HERMIT CRAB
Dardanus fucosus Biffar & Provenzano
by J. W. LaTourrette

Dardanus—of the sea, Greek or Latin; *fucosus*—Latin for under false colors. The specific name refers to the nearly identical gross appearance of this species to *D. venosus*. However, black eye patterns quickly differentiate one from the other, as described by the vernacular nomenclature.

Phylum Arthropoda
Class Crustacea
Order Decapoda
Family Paguridae

STAR-EYED HERMIT CRAB
Dardanus venosus (H. Milne Edwards)
by J. W. LaTourrette

Dardanus—from both Latin and Greek, meaning belonging to the water (sea); *venosus*—Latin for veined, alluding to the veinlike pattern on the inner surface of the major claw. Hermit crabs are fascinating to watch and perform an important function in an aquarium as scavengers. I feed only the fish; excess food seldom has time to decay with hermits on the job.

86

PALMATE HERMIT CRAB*
Pagurus impressus (Benedict)
by J. W. LaTourrette

Pagurus—from a Greek word for a crab; *impressus*—Latin for impressed, referring to the impression on the major claws that serves in gross identification.

* Common name suggested by the author.

Phylum Arthropoda
Class Crustacea
Order Decapoda
Family Paguridae

LONG-ARMED HERMIT CRAB*
Pagurus longicarpus Say
by J. W. LaTourrette

Pagurus—from a Greek word for crab; *longi-carpus*—Latin for long and the wrist joint (long-armed). The crab shown here is housed in a crown conch shell (*Melongena corona*).

* Common name suggested by the author.

87

GRAY HERMIT CRAB*
Pagurus pollicaris Say
by J. W. LaTourrette

Pagurus—from a Greek word for a crab; *pollicaris*—a Latin term for the thumb, probably alluding to the large claw. An anemone (*Calliactus tricolor*) adheres to the shell home of this crab. The species is abundant along the northeast coast of Florida.

* Common name suggested by the author.

Phylum Arthropoda
Class Crustacea
Order Decapoda
Family Paguridae

OPERCULATE HERMIT CRAB*
Pylopagurus operculatus Stimpson
by J. W. LaTourrette

Pylopagurus—derived from the Greek for gate and a crab; *operculatus*—having the nature of an operculum, a lid or cover, Latin. As implied in both the generic and specific nomenclature, the major claw serves as a trap door, or operculum, enabling the crab to seal itself within the shell.

* Common name suggested by the author.

Phylum Arthropoda
Class Crustacea
Order Decapoda
Family Raninidae

88

FROG CRAB*
Ranilia muricata H. Milne Edwards
by Don Renn

Ranilia—an allusion to the Greek word *rana*, frog; *muricata*—spiny in Latin. The specific name refers to the spiny margin along the anterior edge of the carapace. These crabs are most often dredged from offshore sandy bottoms. The family is distinguished by the elongate carapace and dorsally visible abdominal plates (tegra). The photograph does not show the flattened major claws on the first legs, which are tucked under the body.

* Common name suggested by the author.

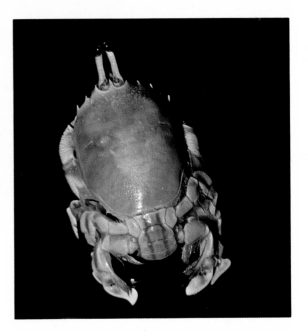

Phylum Arthropoda
Class Crustacea
Order Decapoda
Family Dromiidae

DECORATOR CRAB
Dromidia antillensis Stimpson
by Mike Davis

Dromidia—running or swift in Greek; *antillensis*—a combination of Greek and Latin for horny, alluding to the rigid exoskeleton. In this interesting example of symbiosis, the crab carries a living sponge, ascidians, or zoanthid polyps over its back for camouflage, held securely in place by the specialized fifth pair of legs. The camouflage, unable to survive the plankton-free aquarium environment, is discarded by the crab. A ready and acceptable (to *Dromidia*) substitute can be effected with a plastic sponge.

Phylum Arthropoda
Class Crustacea
Order Decapoda
Family Leucosiidae

89

PURSE CRAB
Persephona mediterrania (Herbst)
by Don Renn

Persephona—the wife of Hades in Greek myth; *mediterrania*—Mediterranean. The common nomenclature was derived from the large purselike receptacle formed by the abdominal segments in which the female carries her sponge of eggs.

Phylum Arthropoda
Class Crustacea
Order Decapoda
Family Calappidae

BOX CRAB
Calappa flammea (Herbst)
by Don Renn

Calappa—an invented name (in new Latin) for the genus; *flammea*—from Latin for flame colored or patterned. Large claws and slender legs neatly fit into corresponding body depressions, eliminating projections that might be seized by a predator. Gastropods, whose shells the box crab can easily break with its powerful claws, constitute a staple diet. The species is inedible!

YELLOW BOX CRAB
Calappa gallus (Herbst)
by J. W. LaTourrette

Calappa—an invented Latin name; *gallus*—fowllike in Latin, in allusion to the upper margin of the claws, which resembles the comb of a cock. As is the case with other family members, this species is inedible.

Phylum Arthropoda
Class Crustacea
Order Decapoda
Family Calappidae

SHAMEFACED CRAB
Hepatus epheliticus (Linnaeus)
by Don Renn

Hepatus—Greek, of or pertaining to the liver, in this case probably a color reference; *epheliticus*—freckled in Greek. Also aptly called calico crab. Were common names based only on appearance, the choice between that apparent expression and color pattern would be difficult.

Phylum Arthropoda
Class Crustacea
Order Decapoda
Family Portunidae

SARGASSUM CRAB
Portunus sayi (Gibbes)
by J. W. LaTourrette

Portunus—Roman god of the port or harbor; *sayi*—a patronymic name. One of the pelagic swimming crabs found among floating sargassum weed. The female shown here bears a sponge of eggs.

Phylum Arthropoda
Class Crustacea
Order Decapoda
Family Portunidae

SWIMMING CRAB*
Portunus spinimanus Latrielle
by J. W. LaTourrette

Portunus—Roman god of the port or harbor; *spinimanus*—Greek for spiny hands. An abundant species, but small in size; thus they are not of commercial importance.

* Common name suggested by the author.

92

BLUE CRAB
Callinectes sapidus Rathbun
by Don Renn

Callinectes—from the Latin *callos*, hard skin, and the Greek word for swimmer; *sapidus*—savory in Latin. This species supports the largest crab-fishing industry in the United States. After molting, it is sometimes unrecognized by consumers as soft-shelled crab. The sexes are easily distinguished in crabs by the shape of the abdomen: that of the male is long and narrow, that of the female triangular or semicircular.

Phylum Arthropoda
Class Crustacea
Order Decapoda
Family Xanthidae

HAIRY CRAB
Pilumnus sayi Rathbun
by J. W. LaTourrette

Pilumnus—Greek for hairy; *sayi*—a patronym. A small xanthid crab found among the rocks in shallow waters.

WARTY CRAB*
Eriphia gonagra (Fabricius)
by J. W. LaTourrette

Eriphia—a Latin or Greek name for an unknown plant; *gonagra*—a Greek word for catching or hunting for prey. Contrasting colors and neat rows of tubercles are distinctive. Turning over stones just above the waterline may reveal these crabs. They have proved hardy in the aquarium.

* Common name suggested by the author.

Phylum Arthropoda
Class Crustacea
Order Decapoda
Family Xanthidae

STONE CRAB
Menippe mercenaria (Say)
by J. W. LaTourrette

Menippe—Greek, meaning force or courage; *mercenaria*— a Latin term referring to something of value. The specific name must allude to the commercial value of this species: the massive claw meat is considered a delicacy. Thus this crab is protected by laws: (a) it may be taken by traps; (b) no spears, hooks, gigs, or grabs are permitted; (c) no females can be taken; (d) only one claw can be removed, and it must measure 4 inches from tip to first joint. The crab must be returned to the water where it can regenerate a new claw. In Florida the closed season is from 1200 hr June 1 to 1200 hr. October 14.

94

BAT CRAB
Carpilius corallinus (Herbst)
by Peter Vila

Carpilius—from the Latin words *carpus*, wrist joint, and *pilus*, hairy; *corallinus*— Latin for coralline in allusion to its coral-red color. The species also bears the common names "coral crab" and "queen crab." It is the largest of West Indian crabs and is often utilized for food.

Phylum Arthropoda
Class Crustacea
Order Decapoda
Family Xanthidae

ROUGH CRAB*
Glyptoxanthus erosus (Stimpson)
by Peter Vila

Glyptoxanthus—from two Greek words, one for carved and the other for yellow; *erosus*—eroded in Latin. Although the shell appears to be furrowed by erosion, the elevated portions are formed by masses of small granules crowded together. In older specimens the shell may be worn smooth.

* Common name suggested by the author.

95

Phylum Arthropoda
Class Crustacea
Order Decapoda
Family Grapsidae

BEACH CRAB
Sesarma ricordi H. Milne Edwards
by J. W. LaTourrette

Sesarma—Greek for moth and Latin for shoulder; *ricordi*—a patronym. The generic name may describe the flighty, mothlike manner in which this small crab scurries about rocks and beach.

Phylum Arthropoda
Class Crustacea
Order Decapoda
Family Grapsidae

MOTTLED SHORE CRAB
Grapsus grapsus Linnaeus
Top by Peter Vila; bottom by J. W. LaTourrette

Grapsus—both generic and specific names are derived from a Greek word for crab. This is a highly terrestrial crab and is most often seen skittering along the shore or among the mangroves. The amazing speed with which it moves prompted its alternative common name, "Suzy lightfoot."

96

Phylum Arthropoda
Class Crustacea
Order Decapoda
Family Ocypodidae

GHOST CRAB
Ocypode quadrata (Fabricius)
by J. W. LaTourrette

Ocypode—Greek for swift-footed (fleet); *quadrata*—
quadrate, or squared, in Latin, alluding to body
form. Quick work with a shovel will reveal the ghost crab
as maker of the mysterious burrows in the sand above the
surf line. A receding tide finds them foraging the debris
left on shore.

97

Phylum Arthropoda
Class Crustacea
Order Decapoda
Family Majidae

ARROW CRAB
Stenorhynchus seticornis (Herbst)
by Peter Vila

Stenorhynchus—a narrow snout or nose in Greek;
seticornis—Latin for bristly horn. An elongated
rostrum and spidery legs make this interesting
small crab an aquarium favorite. This is a rather
aggressive crab and more than once has taken
small prized fish in the same aquarium for a
gourmet meal.

Phylum Arthropoda
Class Crustacea
Order Decapoda
Family Majidae

SPIDER CRAB
Libinia dubia H. Milne Edwards
by Don Renn

Libinia—a patronymic name; *dubia*—Latin for dubious. Larger spider crabs are almost always free of growths; younger specimens frequently are covered with sponges and the like. The adult shown here is a male, as indicated by the elongated narrow abdominal segments seen in the ventral view. Spider crabs living in some form of symbiosis with jellyfishes have been recorded. Most often, however, *Libinia* dwells on almost every type of bottom environment.

98

Phylum Arthropoda
Class Crustacea
Order Decapoda
Family Majidae

SPINY MITHRAX*
Mithrax spinosissimus (Lamarck)
by J. W. LaTourrette

Mithrax—a Latin name for a Persian precious stone; *spinosissimus*—Latin for very thorny or spiny. This is the largest species of the genus. It is found in rocky areas, and the shell may be covered with a variety of growths.

* Common name suggested by the author.

99

CORAL CRAB
Mithrax hispidus (Herbst)
by J. W. LaTourrette

Mithrax—a Latin term referring to a precious stone from Persia; *hispidus*—spiny, also from Latin. The major claws of the female are identical in size; males bear one (referred to as the fighting claw) larger than the other.

Phylum Arthropoda
Class Crustacea
Order Decapoda
Family Majidae

PINCER CRAB*
Mithrax forceps (H. Milne Edwards)
by Peter Vila

Mithrax—a Persian precious stone; *forceps*—
pincers. Both the generic and specific names are
derived from Latin. The specimen, 1 inch in
diameter, emerged from a rock lifted from a
Biscayne Bay breakwater.

* Common name suggested by the author.

MACE CRAB*
Macrocoeloma camptocerum (Stimpson)
by J. W. LaTourrette

Macrocoeloma—from the Greek for large hollow;
camptocerum—Greek for flexible and Latin for
waxen, probably in reference to surface growths.
Growths supported on their carapaces are in keeping with the individual's particular environment.

* Common name suggested by the author.

100

Phylum Arthropoda
Class Crustacea
Order Decapoda
Family Parthenopidae

CROSSCUT CRAB*
Parthenope serrata (H. Milne Edwards)
by J. W. LaTourrette

Parthenope—a siren of Greek myth, said to have been cast
upon the shore of Naples; *serrata*—Latin for serrated.

* So distinctive a dweller of the coastal shallows should not be
without a common name; thus I have taken the liberty of pro-
posing the name "crosscut crab" in allusion to the sawlike
chelipeds.

PHYLUM ECHINODERMATA
THE ECHINODERMS

On seeing a sea star or urchin it is hard to believe that its larval stage displays bilateral symmetry. However, such is the case. As the larvae develop, right and left are forsaken for a five-sectioned, modified radially symmetrical form well adapted to its role in the marine environment. Seemingly flaccid bodies are both protected and supported by skeletons of calcium carbonate plates or granules embedded in the skin. Unlike vertebrate forms, this skeleton is not coordinated with musculature. Locomotion is instead accomplished by a water vascular system, a sort of hydraulically activated movement unique to this almost exclusively marine phylum.

The phylum Echinodermata consists of five classes. Crinoidea encompasses the delicate and lovely sea lillies and feather stars, almost all of which are deep-water forms. Species of these from the southern Atlantic have not been maintained at the Seaquarium. One was on display in a closed-system aquarium at the Rosenstiel School of Marine and Atmospheric Science and did well for some months. The trick seems to be solely in acquiring healthy specimens from the depths. Asteroidea are the sea stars, or "starfishes" as they are often called. Ophiuroidea are the brittle stars. Echinoidea is the class containing sea urchins and sand dollars. Sea cucumbers are within the class Holothuroidea.

Echinoderms are of limited commercial value; studies on the development of new drugs from the sea may prove them to be of greater importance in the future. At present their worth seems limited to preserved curiosities. Few other than holothurians tempt the human palate.

Phylum Echinodermata
Class Asteroidea
Order Phanerozonia
Family Astropectinidae

ROYAL SEA STAR*
Astropecten articulatus (Say)
by Peter Vila

Astropecten—Greek for star and Latin for comb or rake; *articulatus*—Latin for divided into joints. Royal purple and gold inspired the vernacular. Bright colors are unusual for tropical Atlantic asteroids.

* Common name suggested by the author.

102

SEA STAR
Astropecten duplicatus Gray
Top by J. W. LaTourrette; bottom by Peter Vila

Astropecten—from the Greek word for star and Latin for comb or rake, the latter alluding to the even rows of projections along the perimeter of each arm; *duplicatus*—double in Latin. The sea star is one of the most common Asteroidea in this range. The specimens depicted again demonstrate the fact that pattern and color variations alone do not support species differentiations.

Phylum Echinodermata
Class Asteroidea
Order Phanerozonia
Family Luidiidae

LINED SEA STAR*
Luidia clathrata (Say)
by Peter Vila

Luidia—a patronym; *clathrata*—Latin for latticed. The linear pattern is fairly distinctive within this range. If collected with care, they fare well in captivity. Creeping on or under the surface of the sand, they help maintain the artificial marine environment by feeding on organic waste.

* Common name suggested by the author.

103

WEAK SEA STAR*
Luidia alternata (Say)
by J. W. LaTourrette

Luidia—a patronymic generic name; *alternata*—Latin for alternating, referring to the color pattern. Arms bristling with short spines and a yellowish-cream underside help identify this common species. So far they have been found to survive poorly in aquaria; they seem literally to fall apart at the slightest disturbance. All parts must be removed quickly from the aquarium, as they putrify rapidly and can destroy years of the hobbyist's work in a very short time.

* Common name suggested by the author.

Phylum Echinodermata
Class Asteroidea
Order Phanerozonia
Family Luidiidae

NINE-ARMED SEA STAR
Luidia senegalensis (Lamarck)
by Mike Davis

Luidia—a patronymic name; *senegalensis*—of Senegal.
This species has not been found in Florida waters, but can
grow to 1 foot in diameter in its Bahamian range.

Phylum Echinodermata
Class Asteroidea
Order Phanerozonia
Family Oreasteridae

104

CUSHIONED STAR
Oreaster reticulatus (Linnaeus)
by J. W. LaTourrette

Oreaster—possibly derived from a combination of Anglo-
Saxon and Greek words meaning copper or bronze star,
or from Greek and Latin words for little mountain; *reticu-latus*—Latin for netlike (pattern on the body). The
younger mottled green specimen will grow to the red adult.
The cushioned star has proved to be the most hardy of
the sea stars in the aquarium.

Phylum Echinodermata
Class Asteroidea
Order Spinulosa
Family Echinasteridae

SMALL-SPINE SEA STAR*
Echinaster spinulosus Verrill
by Peter Vila

Echinaster—a combination of two Greek words for spiny star; *spinulosus*—Latin for spiny. The nomenclature refers to the multitude of small orange (in this specimen) bumps.

* Common name suggested by the author.

105

SPINY SEA STAR
Echinaster sentus (Say)
by J. W. LaTourrette

Echinaster—Greek for spiny, as in *Echinodermata*, and star; *sentus*—Latin for thorny. All starfishes are infamous for the havoc they wreak in feeding on mollusks. In open-system aquariums they can be utilized as housekeepers; they will feed on almost anything left over by other creatures therein. Their eggs, laid in endless numbers, serve as food for other specimens. Those that survive add to the display as minute mirror images of the adult. When alarmed, starfishes exude a cloudy toxic "mist" into the water from a pore located at the base of the **V** formed between the arms. This will kill almost all life in a recirculating system.

Phylum Echinodermata
Class Ophuiroidea
Order Phrynophiurida
Family Ophiomyxidae

SLIMY BRITTLE STAR*
Ophiomyxa flaccida (Say)
by J. W. LaTourrette

Ophiomyxa—from the Greek words for snake and slime (slime snake); *flaccida*—Latin for flaccid or soft. The color patterns of this species are variable. However, it is the only brittle star within this area that is slimy to the touch. Others are rather hard or scaly.

* Common name suggested by the author.

Phylum Echinodermata
Class Ophuiroidea
Order Phrynophiurida
Family Gorgonocephalidae

106

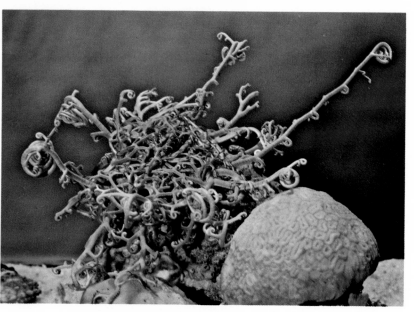

BASKET STAR
Astrophyton muricatum (Lamarck)
by Peter Vila

Astrophyton—from two Greek words meaning star plant; *muricatum*—spiny or pointed in Latin. Five multibranched arms open at night to form an efficient net for ensnaring planktonic food. In an aquarium fish blood, brine shrimp, and macerated foods will serve for a time. More often than not, the specimen dismembers itself into a mass of wiggling parts and is lost.

Phylum Echinodermata
Class Ophuiroidea
Order Chilophiurida
Family Ophiodermatidae

BLACK BRITTLE STAR
Ophiocoma echinata (Lamarck)
by J. W. LaTourrette

Ophiocoma—Greek words meaning hairy serpent; *echinata*—of the order of sea urchins. Spinous or hairy arms are distinctive. This species is similar to the less common O. *riisei* Lütken, which bears sharper spines and red-brown irregular markings on the disk in younger specimens.

Phylum Echinodermata
Class Ophuiroidea
Order Chilophiurida
Family Ophiodermatidae

HARLEQUIN BRITTLE STAR*
Ophioderma appressum (Say)
by J. W. LaTourrette

Ophioderma—reptile or snake skin in Greek; *appressum*—Latin for pressed down, or flattened. The patterns on the disk are variable in this common brittle star. It is quite hardy and an interesting addition to any balanced aquarium.

* Common name suggested by the author.

Phylum Echinodermata
Class Ophuiroidea
Order Chilophiurida
Family Ophiocomidae

SPOTTED BRITTLE STAR*
Ophioderma guttatum Lütken
by J. W. LaTourrette

Ophioderma—Greek for reptile or snake skin; *guttatum*—spotted in Latin. The insert shows the ventral side of this rather rare member of the genus.

* Common name suggested by the author.

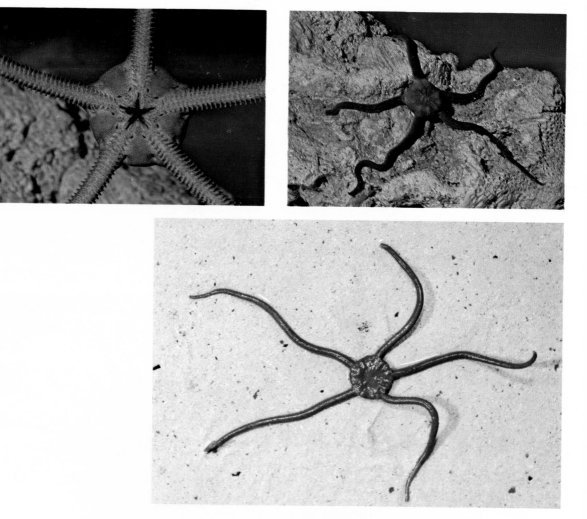

108

SCALY BRITTLE STAR*
Ophioderma squamosissimum Lütken
by J. W. LaTourrette

Ophioderma—Greek for reptile or snake skin; *squamosissimum*—very scaly in Latin. Severed arms are lost; if they are removed with a portion of the central disk, regeneration will yield two complete individuals.

* Common name suggested by the author.

Phylum Echinodermata
Class Echinoidea
Order Cidaroida
Family Cidaridae

CLUB URCHIN
Eucidaris tribuloides (Lamarck)
by Don Renn

Eucidaris—from a Greek term for original or primitive and Latin for a Persian crown; *tribuloides*—Latin for thorny. This shallow-water form is most often found wedged among rocks near the low-tide level. Blunt spines enable it to be handled with ease.

Phylum Echinodermata
Class Echinoidea
Order Centrechinoida
Family Centrechinidae

109

LONG-SPINED SEA URCHIN
Diadema antillarum Philippi
by Peter Vila

Diadema—crown in Latin; *antillarum*—of the Antilles. The spines are coated with a mucilaginous irritant poison; they are hollow, with whorls of microscopic spinelets pointing distally. Although it is apparently sightless, shadows falling across light-sensitive areas stimulate responsive movement of the spines toward that section for defense.

Phylum Echinodermata
Class Echinoidea
Order Centrechinoida
Family Arbaciidae

PURPLE-SPINED SEA URCHIN*
Arbacia punctulata (Lamarck)
by Peter Vila

Arbacia—from the Greek name Arbakés, the first king of Media, perhaps in reference to a royal crown; *punctulata*—punctured in Latin. What specific name could be more appropriate for so formidable a defense as that borne by this echinoderm!

* Common name suggested by the author.

Phylum Echinodermata
Class Echinoidea
Order Centrechinoida
Family Echinidae

GREEN URCHIN
Lytechinus variegatus (Leske)
Left by Peter Vila; right by Mike Davis

Lytechinus—from the Greek for broken and sea urchin; *variegatus*—variegated in Latin. A shallow-water form found on sandy bottoms supporting growths of eel or turtle grasses. A more apt common name might be "variable urchin," as manifested by the two divergent color forms shown here.

Phylum Echinodermata
Class Echinoidea
Order Centrechinoida
Family Echinometridae

ROCK URCHIN

Echinometra lucunter (Linnaeus)
by Don Renn

Echinometra—from two Grecian terms referring to sea urchins and an entrance to the womb; *lucunter*—a kind of cake in Latin. These mainly are shallow-water forms that cling to rocks with tiny tube feet visible above. Their boring ability enables them to modify, if not excavate, suitably protective cavities on stony surfaces in the surf zone.

Phylum Echinodermata
Class Echinoidea
Order Exocycloida
Family Clypeasteridae

SAND DOLLAR

Clypeaster subdepressus (Gray)
by Peter Vila

Clypeaster—a star-bearing shield, from two Latin words; *subdepressus*—also from Latin words descriptive of a ventral depression. Unlike *Encope* (the notched sand dollar), this species bears no marginal notches and is noticeably raised in the center, rather than flat.

Phylum Echinodermata
Class Echinoidea
Order Exocycloida
Family Clypeasteridae

SEA BISCUIT
Clypeaster rosaceus (Linnaeus)
Live specimen by J. W. LaTourrette; bleached test by Don Renn

Clypeaster—from two Latin words for shield and star; *rosaceus*—Latin for rosy, reddish in color. Most often found below the surface of the sand, this echinoderm achieves mobility via movement of tiny species covering the shell. The bleached test displays the five-rayed form typical of echinoderms.

112

Phylum Echinodermata
Class Echinoidea
Order Exocycloida
Family Scutellidae

NOTCHED SAND DOLLAR
Encope michelini Agassiz
Live specimen by J. W. LaTourrette; bleached test by
Don Renn

Encope—from the Greek for cut in; *michelini*—a patronym. The generic name alludes to the marginal notches. A near relative, *Mellita testudinata*, bears notches that are closed along the margin; hence the old vernacular name "key-hole urchin." Tiny spines serve for locomotion and hold sand and debris for camouflage.

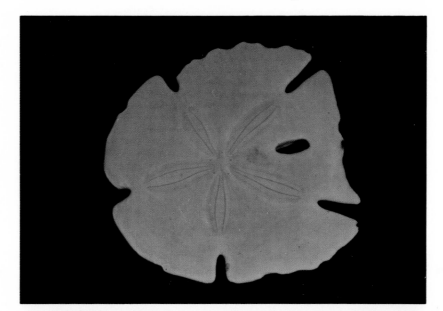

113

Phylum Echinodermata
Class Holothuroidea
Order Aspidochirota
Family Holothuriidae

WEST INDIAN SEA CUCUMBER
Actinopyga agassizi Selenka
by Mike Davis

Actinopyga—Greek for rayed and rump or bottom; *agassizi*—a patronym. A fairly common form inhabiting eel-grass flats or found under the shelter of rock or coral fragments on reef flats. Many individuals are inhabited by small commensal fish (the pearlfish).

114

FISSURED SEA CUCUMBER*
Astichopus multifidus (Sluiter)
by J. W. LaTourrette

Astichopus—Greek for not *Stichopus*, another genus, whose name is taken from the Greek for line or row and refers to the longitudinal bands of pedicels, and a Latin word for discharge of corrupt matter; *multifidus*—many-cleft in Latin.

* Common name suggested by the author.

Phylum Echinodermata
Class Holothuroidea
Order Aspidochirota
Family Holothuriidae

SEA CUCUMBER
Isostichopus badionotus Selenka
by J. W. LaTourrette

Isostichopus—near or like *Stichopus* (see the preceding caption for *Astichopus*); *badionotus*—Latin, to advance slowly. The allusion to discharge of corrupt matter in the generic name probably stems from the fact that, when endangered, these creatures disgorge their entrails. The foul sticky mass is highly toxic to other specimens in the aquarium and, in this contained environment, to the cucumber as well. In the open sea, however, they often survive by regenerating a new set of internal organs.

115

Phylum Echinodermata
Class Holothuroidea
Order Dendrochirota
Family Cucumariidae

PIGMY SEA CUCUMBER*
Pentacta pygmaea (Theel)
by J. W. LaTourrette

Pentacta—Greek for five and Latin for strands or promontories; *pygmaea*—dwarf or little in Greek. As implied by the specific name, this holothurian does not usually achieve more than 3 or 4 inches in length. It is found more commonly along Florida's Gulf Coast than on the eastern shore.

* Common name suggested by the author.

116

BIBLIOGRAPHY

Abbott, R. Tucker, *American Seashells*, D. Van Nostrand Co., Inc., Princeton, N.J., 1954.

Abbott, R. Tucker, *Seashells of North America*, Western Publishing Co., Inc. (Golden Field Guides), New York, 1969.

Amos, William H., *The Life of the Seashore*, McGraw-Hill Book Co., Inc., New York, 1966.

Bayer, Frederick M., "Observations On Pelagic Mollusks Associated with the Siphonophores *Velella* and *Physalia*," *Bulletin of Marine Sciences of the Gulf and Caribbean*, University of Miami Press, Coral Gables, Fla., Vol. 13, No. 3, September 1963, pp. 454–466.

Bayer, Frederick M., *The Shallow Water Octocorallia of the West Indian Region*, Martinus Nijhoff, The Hague, Netherlands, 1961.

Bayer, Frederick M. and Owre, Harding B., *The Free-living Lower Invertebrates*, The Macmillan Co., New York, 1968.

Bennett, Isobel, *The Fringe of the Sea*, Rigby Limited, Adelaide, Australia, 1966.

Buchsbaum, Ralph, *Animals Without Backbones*, Revised Edition, The University of Chicago Press, Chicago, Ill., 1948.

Buchsbaum, Ralph and Milne, Lorus J., *The Lower Animals, Living Invertebrates of the World*, Doubleday & Co., Inc., New York, 1962.

Chester, Richard H., *The Systematics of Sympatric Species of West Indian Spatangoids*, University of Miami Press, Coral Gables, Fla., 1968.

Clark, Hubert Lyman, *A Handbook of the Littoral Echinoderms of Porto Rico and Other West Indian Islands*, Scientific Survey of Puerto Rico and the Virgin Islands, Vol. XVI, Part 1, New York Academy of Sciences, 1933.

Coe, Wesley Roswell, *Starfishes, Serpentstars, Sea Urchins and Sea Cucumbers of the Northeast*, Dover Publications, New York, 1972.

Dales, R. Phillips, *Annelids,* Hutchinson University Library, Hutchinson & Company, Ltd., London, 1963.

Gosner, Kenneth L., *Guide to Identification of Marine and Estuarine Invertebrates—Cape Hatteras to the Bay of Fundy,* John Wiley and Sons, Inc., New York, 1971.

Halstead, Bruce, *Poisonous and Venomous Marine Animals of the World,* Vol. 1, U.S. Government Printing Office, Washington, D.C., 1965.

Holthuis, Lipke B., *A General Revision of the Palaemonidae (Crustacea Decapoda Natantia) of the Americas,* Part 1, Allan Hancock Foundation Publications of the University of Southern California, University of Southern California Press, Los Angeles, 1951.

Hyman, Libbie Henrietta, *The Invertebrates: Protozoa Through Ctenophora,* Vol. 1, McGraw-Hill Book Co., Inc., New York, 1940.

Jaeger, Edmund C., *A Source Book of Biological Names and Terms,* 3rd Edition, Charles C Thomas, Springfield, Ill., 1959.

Johnsonia, Monographs of the Mollusks of the Western Atlantic, William J. Clench, Editor, Published by the Department of Mollusks, Museum of Comparative Zoology, Harvard University, Cambridge, Mass. Vol. 1, 1941–1945; Vol. II, 1946–1953; Vol. III, 1954–1959; Vol. IV, 1960–   (additional volumes available).

King, J., "Recirculating System Culture Methods," in *Invertebrate Culture,* Plenum Publishing Co., New York, 1972.

Klingel, Gilbert C., *Ocean Island,* Doubleday and Co., Inc., New York, 1961.

Limbaugh, Conrad, Pederson, Harry, and Chase, Fenner A., Jr., "Shrimps That Clean Fishes," *Bulletin of Marine Sciences of the Gulf and Caribbean,* University of Miami Press, Coral Gables, Fla., Vol. 11, No. 2, June 1961, pp. 237–257.

Manning, Raymond B., *Stomatopod Crustacea of the Western Atlantic,* University of Miami Press, Coral Gables, Florida.

Marcus, Eveline and Marcus, Ernst, *American Opisthobranch Mollusks,* Studies in Tropical Oceanography No. 6, University of Miami, Institute of Marine Sciences, Miami, Fla., 1967.

Mayer, Alfred Goldsborough, *Medusae of the World,* Vol. III, Carnegie Institute of Washington, Publication 109, Washington, D.C., 1910 (additional volumes available).

Milne, Lorus and Milne, Margery, *Invertebrates of North America*, Doubleday and Co., Inc., New York, 1972.

Minor, Roy Waldo, *Field Book of Seashore Life*, G. P. Putnam's Sons, New York, 1950.

Moore, Hilary B., *Marine Ecology*, John Wiley & Sons, Inc., New York, 1966.

Morton, J. B., *Molluscs*, Harper and Brothers, New York, 1960.

Physiology of Echinodermata, Richard A. Boolootian, Editor, John Wiley & Sons, Inc., New York, 1966.

Pilsbury, H. A., "The Barnacles (Cirripedia) Contained in the Collections of the U.S. National Museum," *U.S. National Museum Bulletin 60*, Washington, D.C., 1907.

Rathbun, Mary J., *The Crancoid Crabs of America*, Smithsonian Institute, Bulletin 152, U.S. Government Printing Office, Washington, D.C., 1930.

Schmitt, Waldo L., *Crustaceans*, University of Michigan Press, Ann Arbor, 1965.

Smith, F. G. Walton, *Atlantic Reef Corals* (Revised), University of Miami Press, Coral Gables, Fla., 1971.

Spotte, S., *Marine Aquarium Keeping, the Science, the Animals, and the Art*, Wiley-Interscience, New York, 1973.

U.S. Fish Commission, *Investigation of the Aquatic Resources and Fisheries of Porto Rico, Bulletin of the United States Fish Commission*, Vol. XX for 1900, U.S. Government Printing Office, Washington, D.C., 1902.

Voss, Gilbert L., "A Review of the Cephalopods of the Gulf of Mexico," *Bulletin of Marine Sciences of the Gulf and Caribbean*, University of Miami Press, Coral Gables, Fla., Vol. 6, No. 2, June 1956, pp. 85–178.

Warmke, G. L. and Abbott, R. Tucker, *Caribbean Sea Shells*, Livingston Publishing Co., Wynnewood, Pa., 1961.

Williams, Austin B., *Marine Decapod Crustaceans of the Carolinas, Fishery Bulletin*, Vol. 65, No. 1, U.S. Fish and Wildlife Service, Washington, D.C., 1965.

APPENDIX

All the specimens presented in this book are listed in phyletic sequence (from more primitive to more specialized life forms) in the table that follows.

PHYLUM	CLASS	ORDER	FAMILY	GENUS	SPECIES	COMMON NAME
Coelenterata	Hydrozoa	Hydrocorallinae	Milleporidae	Millepora	alcicornis	Fire coral
					complanata	Stinging coral
		Siphonophora	Chondrophoridae	Physalia	pelagica	Portuguese man-of-war
				Porpita	umbella	Porpita
				Velella	velella	By-the-wind sailor
	Scyphozoa	Rhizostomae	Cassiopeidae	Cassiopea	xamachana	Jamaican cassiopea*
		Semaeostomeae	Pelagidae	Chrysaora	quinquecirrha	Common sea nettle
	Anthozoa	Zoanthidea	Zoanthidae	Zoanthus	sociatus	Green sea mat
		Actinaria	Aliciidae	Lebrunia	danae	Stinging anemone
			Actiniidae	Actinia	bermudensis	Maroon anemone
				Condylactis	gigantea	Sea anemone
			Stoichactiidae	Stoichactis	helianthus	Sun anemone
			Phymanthidae	Phymanthus	crucifer	Red beaded anemone
			Hormathiidae	Calliactus	tricolor	Hermit anemone*
			Aiptasiidae	Aiptasia	pallida	Pale sea anemone
				Bartholomea	annulata	Ringed anemone*
		Scleractinia	Acroporidae	Acropora	cervicornis	Staghorn coral
					palmata	Elkhorn coral
			Poritidae	Porites	astreoides	Yellow porites*
					porites	Clubbed finger coral
			Faviidae	Diplora	labyrinthiformis	Brain coral
					strigosa	Common brain coral
				Manicina	areolata	Common rose coral
				Montastraea	cavernosa	Large star coral
			Oculinidae	Oculina	diffusa	Ivory bush coral
			Trochosmiliidae	Meandrina	meandrites	Brain coral
				Dichocoenia	stokesi	Domed star coral*
				Dendrogyra	cylindricus	Pillar coral
			Mussidae	Mussa	angulosa	Large flower coral
			Caryophillidae	Eusmilia	fastigiata	Flower coral
		Corallimorpharia	Actinodiscidae	Ricordia	florida	False coral
				Rhodactis	sanctithomae	Red false coral*
Platyhelminthes	Turbellaria	Polycladida	Pseudoceridae	Pseudoceros	sp.	Polyclad flatworm
					crozieri	Crozier's flatworm*
Mollusca	Gastropoda	Archaeogastropoda	Fissurellidae	Diodora	cayenensis	Cayenne keyhole limpet
				Lucapina	suffusa	Cancellate fleshy limpet
			Turbinidae	Astraea	caelata	Carved star shell
			Neritidae	Nerita	peloronta	Bleeding tooth

(continued)

121

122

PHYLUM	CLASS	ORDER	FAMILY	GENUS	SPECIES	COMMON NAME
Mollusca (continued)	Gastropoda (continued)	Archaeogastropoda (continued)	Janthinidae	Janthina	janthina	Common purple sea snail
			Strombidae	Strombus	alatus	Florida fighting conch
					gigas	Queen conch
			Cypraeidae	Cypraea	cervus	Atlantic deer cowrie
					zebra	Measled cowrie
			Ovulidae	Cyphoma	gibbosum	Flamingo tongue
			Cassididae	Phalium	granulatum	Scotch bonnet
				Cassis	flammea	Flame helmet
					madagascariensis	Emperor helmet
					madagascariensis spinella	Clench's helmet
		Neogastropoda	Ficidae	Ficus	communis	Common fig shell
			Muricidae	Murex	pomum	Apple murex
					fulvescens	Giant eastern murex
				Thais	deltoidea	Deltoid rock shell
			Melongenidae	Busycon	contrarium	Lightning whelk
			Fasciolariidae	Fasciolaria	tulipa	True tulip
				Pleuroploca	gigantea	Horse conch
			Olividae	Oliva	sayana	Lettered olive
			Marginellidae	Prunum	apicinum	Common Atlantic marginella
		Tectibranchia	Bullidae	Bulla	occidentalis	Common West Indian bubble
		Ascoglossa	Elysiidae	Tridachia	crispata	Ribbon nudibranch*
		Anaspidea	Aplysiidae	Aplysia	dactylomela	Spotted sea hare
					morio	Black sea hare
				Dolabrifera	dolabrifera	Green sea hare*
				Bursatella	leachi plei	Ragged sea hare
		Notaspidea	Pleurobranchidae	Pleurobranchus	areolatus	Nubby sea slug*
		Doridoidea	Polyceridae	Polycera	hummi	Horned nudibranch*
			Dorididae	Felimare	bayeri	Sea cat nudibranch*
				Hypselodoris	edenticulata	Greek goddess*
				Peltodoris	greeleyi	Greeley's nudibranch*
				Platydoris	angustipes	Leathery nudibranch*
		Eolidoidea	Fionidae	Fiona	pinnata	Feathered nudibranch*
			Facelinidae	Learchis	poica	Orange-lined nudibranch
			Scyllaeidae	Scyllaea	pelagica	Sargassum nudibranch
			Aeolidiidae	Spurilla	neapolitana	Spurred nudibranch*
			Glaucidae	Glaucus	marinus	Blue glaucus

PHYLUM	CLASS	ORDER	FAMILY	GENUS	SPECIES	COMMON NAME
Mollusca (continued)	Amphineura	Chitonida	Chitonidae	Acanthopleura	granulata	Fuzzy chiton
	Pelecypoda	Filibranchia	Pectinidae	Pecten	ziczac	Zigzag scallop
			Spondylidae	Spondylus	americanus	Atlantic thorny oyster
			Limidae	Lima	scabra tenera	Delicate lima*
				Lima	scabra	Rough lima
			Pholadidae	Martesia	cuneiformis	Wedge-shaped martesia
			Gastrochaenidae	Spengleria	rostrata	Spengler clam
	Cephalopoda	Decapoda	Loliginidae	Sepioteuthis	sepioidea	Atlantic oval squid
			Ommastrephidae	Ommastrephes	pteropus	Orange-back squid
		Octopoda	Argonautidae	Argonauta	argo	Paper nautilus
			Octopodidae	Octopus	briareus	Briar octopus
				Octopus	joubini	Dwarf octopus
				Octopus	macropus	White spotted octopus
				Octopus	vulgaris	Common Atlantic octopus
Annelida	Polychaeta	Sabellida	Sabellidae	Sabella	melanostigma	Black-spotted sabellid worm*
			Serpulidae	Serpulid†	sp.	Serpulid worm
		Amphinomida	Amphinomidae	Chloeia	viridis	Bristle worm
Arthropoda	Arachnomorpha	Merostomata	Limulidae	Limulus	polyphemus	Horseshoe crab
	Crustacea	Cirripedia	Balanidae	Chelonibia	patula	Cloven barnacle*
			Lepadidae	Lepas	ansifera	Gooseneck barnacle
		Stomatopoda	Lysiosquillidae	Lysiosquilla	scabricauda	Mantis shrimp
			Gonodactylidae	Pseudosquilla	ciliata	False mantis shrimp* shrimp
			Squillidae	Squilla	rugosa	Armored mantis
		Decapoda	Penaeidae	Penaeus	duorarum	Pink shrimp
				Sicyonia	brevirostris	Rock shrimp
			Palaemonidae	Periclimenes	parri	Parr's rock shrimp*
					petersoni	Peterson shrimp
					yucatanicus	Spotted cleaner shrimp
				Brachycarpus	biunguiculatus	Two-clawed shrimp*
				Pontonia	mexicana	Pen shrimp*
			Stenopidae	Microprosthema	semilaevis	Lima shrimp*
				Stenopus	hispidus	Banded coral shrimp
					scutellatus	Golden coral shrimp
			Gnathophyllidae	Gnathophyllum	americanum	Bumble bee shrimp
			Alpheidae	Alpheus	armillatus	Banded snapping shrimp

(continued)

PHYLUM	CLASS	ORDER	FAMILY	GENUS	SPECIES	COMMON NAME
Arthropoda (continued)	Crustacea (continued)	Decapoda (continued)	Alpheidae (continued)	Alpheus (continued)	*formosus*	Striped snapping shrimp
					brevicarpus	Snapping shrimp
				Synalpheus	*grabhami*	Scarlet lady*
			Hippolytidae	*Hippolysmata*	*wurdemanni*	Veined shrimp*
			Homaridae	*Homarus*	*americanus*	American lobster
			Palinuridae	*Panulirus*	*argus*	Spiny lobster
					guttatus	Rock lobster
					laevicauda	Brazilian lobster
				Palinurellus	*gundlachi*	Copper lobster*
				Justitia	*longimanus*	Long-armed lobster*
			Scyllaridae	*Scyllarus*	*americanus*	Spanish lobster
				Scyllarides	*aequinoctalis*	Locust lobster
			Axiidae	*Axiopsis*	sp.	Burrowing shrimp*
			Porcellanidae	*Porcellana*	*sayana*	Spotted porcelain crab
			Coenobitidae	*Coenobita*	*clypeatus*	Land hermit crab
			Paguridae	*Paguristes*	*cadenati*	Red hermit crab*
				Clibanarius	*vittatus*	Striped hermit crab
				Petrochirus	*diogenes*	Hermit crab
				Dardanus	*fucosus*	Bar-eyed hermit crab
					venosus	Star-eyed hermit crab
				Pagurus	*impressus*	Palmate hermit crab*
					longicarpus	Long-armed hermit crab*
				Pylopagurus	*pollicaris*	Gray hermit crab*
					operculatus	Operculate hermit crab*
			Raninidae	*Ranilia*	*muricata*	Frog crab*
			Dromiidae	*Dromidia*	*antillensis*	Decorator crab
			Leucosiidae	*Persephona*	*mediterrania*	Purse crab
			Calappidae	*Calappa*	*flammea*	Box crab
					gallus	Yellow box crab
				Hepatus	*epheliticus*	Shamefaced crab
			Portunidae	*Portunus*	*sayi*	Sargassum crab
					spinimanus	Swimming crab*
				Callinectes	*sapidus*	Blue crab
			Xanthidae	*Pilumnus*	*sayi*	Hairy crab
				Eriphia	*gonagra*	Warty crab*
				Menippe	*mercenaria*	Stone crab
				Carpilius	*corallinus*	Bat crab
				Glyptoxanthus	*erosus*	Rough crab*
				Sesarma	*ricordi*	Beach crab

PHYLUM	CLASS	ORDER	FAMILY	GENUS	SPECIES	COMMON NAME
Arthropoda (continued)	Crustacea (continued)	Decapoda (continued)	Grapsidae	Grapsus	grapsus	Mottled shore crab
			Ocypodidae	Ocypode	quadrata	Ghost crab
			Majidae	Stenorhynchus	seticornis	Arrow crab
				Libinia	dubia	Spider crab
				Mithrax	spinossissimus	Spiny mithrax*
					hispidus	Coral crab
					forceps	Pincer crab*
				Macrocoeloma	camptocerum	Mace crab*
			Parthenopidae	Parthenope	serrata	Crosscut crab*
Echinodermata	Asteroidea	Phanerozonia	Astropectinidae	Astropecten	articulatus	Royal sea star*
					duplicatus	Sea star
			Luididae	Luidia	clathrata	Lined sea star*
					alternata	Weak sea star*
					senegalensis	Nine-armed sea star
		Spinulosa	Oreasteridae	Oreaster	reticulatus	Cushioned star
			Echinasteridae	Echinaster	spinulosus	Small-spine sea star*
					sentus	Spiny sea star
	Ophiuroidea	Phrynophiurida	Ophiomyxidae	Ophiomyxa	flaccida	Slimy brittle star*
			Gorgonocephalidae	Astrophyton	muricatum	Basket star
		Chilophiurida	Ophiocomidae	Ophiocoma	echinata	Black brittle star
			Ophiodermatidae	Ophioderma	appressum	Harlequin brittle star*
					guttatum	Spotted brittle star*
					squamosissimum	Scaly brittle star*
	Echinoidea	Cidaroida	Cidaridae	Eucidaris	tribuloides	Club urchin
		Centrechinoida	Centrechinidae	Dialema	antillarum	Long-spined sea urchin
			Arbaciidae	Arbacia	punctulata	Purple-spined sea urchin*
		Exocycloida	Echinidae	Lytechinus	variegatus	Green urchin
			Echinometridae	Echinometra	lucunter	Rock urchin
			Clypeasteridae	Clypeaster	rosaceus	Sea biscuit
					subdepressus	Sand dollar
			Scutellidae	Mellita	testudinata	Notched sand dollar
	Holothuroidea	Aspidochirota	Holothuriidae	Actinopyga	agassizi	West Indian sea cucumber
				Isostichopus	badionotus	Sea cucumber
				Astichopus	multifidus	Fissured sea cucumber*
		Dendrochirota	Cucumariidae	Pentacta	pygmaea	Pigmy sea cucumber*

* Common name suggested by the author.
† Not a generic name.

INDEX TO COMMON NAMES

INDEX TO SCIENTIFIC NAMES

129

DATE DUE

5-13-7?			
JAN 30			
Jul 27 '82			
GAYLORD			PRINTED IN U.S.A.